Springer Tracts in Modern Physics 89

D1642031

Springer Tracts in Modern Physics

66* **Quantum Statistics in Optics and Solid-State Physics**
With contributions by R. Graham, F. Haake

67* **Conformal Algebra in Space-Time and Operator Product Expansion**
By S. Ferrara, R. Gatto, A. F. Grillo

68* **Solid-State Physics** With contributions by D. Bäuerle, J. Behringer, D. Schmid

69* **Astrophysics** With contributions by G. Börner, J. Stewart, M. Walker

70* **Quantum Statistical Theories of Spontaneous Emission and their Relation to Other Approaches** By G. S. Agarwal

71 **Nuclear Physics** With contributions by J. S. Levinger, P. Singer, H. Überall

72 **Van der Waals Attraction:** Theory of Van der Waals Attraction By D. Langbein

73 **Excitons at High Density** Edited by H. Haken, S. Nikitine. With contributions by V. S. Bagaev, J. Biellmann, A. Bivas, J. Goll, M. Grosmann, J. B. Grun, H. Haken, E. Hanamura, R. Levy, H. Mahr, S. Nikitine, B. V. Novikov, E. I. Rashba, T. M. Rice, A. A. Rogachev, A. Schenzle, K. L. Shaklee

74 **Solid-State Physics** With contributions by G. Bauer, G. Borstel, H. J. Falge, A. Otto

75 **Light Scattering by Phonon-Polaritons** By R. Claus, L. Merten, J. Brandmüller

76 **Irreversible Properties of Type II Superconductors** By. H. Ullmaier

77 **Surface Physics** With contributions by K. Müller, P. Wißmann

78 **Solid-State Physics** With contributions by R. Dornhaus, G. Nimtz, W. Richter

79 **Elementary Particle Physics** With contributions by E. Paul, H. Rollnick, P. Stichel

80* **Neutron Physics** With contributions by L. Koester, A. Steyerl

81 **Point Defects in Metals I:** Introduction to the Theory 2nd Printing
By G. Leibfried, N. Breuer

82 **Electronic Structure of Noble Metals, and Polariton-Mediated Light Scattering**
With contributions by B. Bendow, B. Lengeler

83 **Electroproduction at Low Energy and Hadron Form Factors**
By E. Amaldi, S. P. Fubini, G. Furlan

84 **Collective Ion Acceleration** With contributions by C. L. Olson, U. Schumacher

85 **Solid Surface Physics** With contributions by J. Hölzl, F. K. Schulte, H. Wagner

86 **Electron-Positron Interactions** By B. H. Wiik, G. Wolf

87 **Point Defects in Metals II:** Dynamical Properties and Diffusion Controlled Reactions
With contributions by P. H. Dederichs, K. Schroeder, R. Zeller

88 **Excitation of Plasmons and Interband Transitions by Electrons**
By H. Raether

89 Giant Resonance Phenomena in **Intermediate-Energy Nuclear Reactions**
By F. Cannata, H. Überall

* denotes a volume which contains a Classified Index starting from Volume 36.

F. Cannata H. Überall

Giant Resonance Phenomena in **Intermediate-Energy Nuclear Reactions**

With 43 Figures

Springer-Verlag
Berlin Heidelberg New York 1980

Dr. Francesco Cannata

Istituto di Fisica, Università di Bologna,
INFN Sezione di Bologna, Italy

Professor Dr. Herbert Überall

The Catholic University of America, Dept. of Physics,
Washington, DC 20064, and
Naval Research Laboratory,
Washington, DC 20375, USA

Manuscripts for publication should be addressed to:

Gerhard Höhler

Institut für Theoretische Kernphysik der Universität Karlsruhe
Postfach 6380, D-7500 Karlsruhe 1, Fed. Rep. of Germany

Proofs and all correspondence concerning papers in the process of publication should be addressed to:

Ernst A. Niekisch

Haubourdinstrasse 6, D-5170 Jülich 1, Fed. Rep. of Germany

ISBN 3-540-10105-5 Springer-Verlag Berlin Heidelberg New York
ISBN 0-387-10105-5 Springer-Verlag New York Heidelberg Berlin

Library of Congress Cataloging in Publication Data. Cannata, F. 1945-. Giant resonance phenomena in intermediate-energy nuclear reactions. (Springer tracts in modern physics ; 89). Includes bibliographical references and index. 1. Nuclear magnetic resonance, Giant. 2. Nuclear reactions. I. Überall, Herbert, 1931- joint author. II. Title. III. Series. QC1.S797 vol. 89 [QC762] 530s [538'.362] 80-14938

Offset printing and bookbinding: Brühlsche Universitätsdruckerei, Giessen
2153/3130 — 5 4 3 2 1 0

Preface

In 1947, Baldwin and Klaiber carried out experiments bombarding atomic nuclei with gamma quanta, and observed that at certain high (15-20 MeV) excitation energies, the nucleus began to act as a strong absorber of the incident photons. This phenomenon was termed the "nuclear giant resonance", and was quickly interpreted by Goldhaber and Teller (1948) as the excitation of a collective mode of nuclear vibrations in which all the protons in the nucleus moved together against all the neutrons, thus performing a "giant" dipole vibration. A universal nuclear feature, giant dipole resonances were found to exist in all nuclei throughout the periodic table, originally mainly via photonuclear experiments.

A decisive step forward occurred with the hypothesis (Foldy and Walecka, 1964) that the giant resonances formed isomultiplets, and hence would also be present in neighboring isobars where they could be excited in charge-exchange reactions such as muon capture. This approach succeeded in explaining muon capture rates in terms of photonuclear cross sections.

With the advent of the inelastic electron scattering technique, the nuclear giant resonances were shown to be excited in this type of experiment also. In addition, a new mode of dipole vibrations of nuclear matter was discovered, one in which protons with spin up move collectively, together with neutrons with spin down, against the remaining protons with spin down and neutrons with spin up (Überall, 1965). These spin and isospin extensions of the giant resonances permit their description in terms of an SU4 vector supermultiplet in the framework of Wigner's nuclear supermultiplet theory.

The most recent advance in the field occurred with the discovery of giant resonance vibrations of higher multipolarity, using proton scattering (Lewis and Bertrand, 1972) and electron scattering (Pitthan and Walcher, 1971; Fukuda and Torizuka, 1972). Such higher-order collective multipole vibrations had been theoretically predicted long before (Danos, 1952), and were classified in terms of SU4 multiplets (Raphael, Überall and Werntz, 1966).

The foregoing has already implied that giant resonances, as a property of the nuclear spectrum, may be excited in a variety of nuclear reactions, weak, electromagnetic and strong, employing different types of projectiles and reaction mechanisms. We are thus offered the advantage that this interesting nuclear feature

can be investigated from many different viewpoints, so that taking all these approaches together, we may obtain a quite complete description of it. The present monograph capitalizes on this fact, and discusses our progress in giant resonance studies as obtained through photonuclear and electron scattering reactions, muon capture and neutrino excitations, pion photoproduction and radiative capture, and finally through strong interactions of protons, alpha particles, and pions. We hope that in this way the reader has been presented a fairly complete (although perhaps not exhaustive) picture of our present status of knowledge on the giant nuclear resonances, as well as of the way in which this knowledge has been obtained.
We wish to acknowledge the support of the National Science Foundation in the preparation of this report. One of us (F.C.) is particularly indebted to the late Prof. J.I. Fujita for many enlightening discussions.

Bologna, Italy Francesco Cannata

Bethesda, MD, USA Herbert Überall

April 1980

Contents

Introduction

Giant resonance excitation in nuclei, first observed in 1947 in photonuclear absorption experiments /1/, have been shown to constitute a universal nuclear phenomenon, upon which much subsequent interest has been focused. Previous studies have for a long time been restricted first to the electric dipole components of the resonances (see, e.g., /2, 3/), with later extension to giant magnetic dipole states /4/; but the subject has received renewed interest through the more recent discovery of higher multipole giant resonances (/5/, see also /6-11/), whose existence had been discussed /12/ several years before their actual discovery. In the mentioned theoretical study predicting these resonances /12/, attention had been given to the effects they would produce in inelastic electron scattering experiments /13/; and in fact, giant electric quadrupole resonances were indeed observed in such reactions also /14/, but independently in inelastic proton and alpha scattering experiments as well.

This example illustrates an additional feature of the giant resonances in general, which puts them among the most interesting objects of study for intermediate-energy nuclear experiments: namely, the fact that they can be excited in a variety of nuclear reactions, involving electromagnetic, weak, as well as strong interactions. The advantages of a multi-faceted study of this single nuclear feature which thereby offers itself, are considerable. The information thus gained regarding the nuclear transition densities of the giant resonance states, overlaps in the various reaction channels of the intermediate energy regime, hence providing corroborating information on the level structure. Selection rules in different reactions (such as spin, isospin, etc.) will emphasize one or the other feature and hence provide complementary information on the transition densities not obtainable with a single probe.

In the present review we shall adopt the viewpoint, just outlined, of a multi-reaction approach for the study of giant resonance phenomena. A unified treatment is possible for the latter, because although excitation processes such as neutrino-induced reactions and muon capture have a different SU4 geometry, the amplitudes are the same as, e.g., for electroexcitation processes and photoreactions. The forms of the nuclear interactions for these various processes are discussed in Chap. 1 as well as the classification of the various giant resonance components based on certain nuclear models. This section discusses also the dynamics of the giant resonance based on hydrodynamical models /12/. Their interpretation in terms of Regge poles /15/ is given in Chap. 2. That section also deals with the giant resonance form factors that are studied in inelastic electron scattering. We emphasize the role of spin-flip contributions which are important in electronuclear processes, not only for the well-established spin-isospin giant resonance excitation, but also for their interference with the isospin excitation which is sensitive to the ground state SU4 impurities of the target. There exists a similar situation in muon capture, the subject of Chap. 3, for which a microscopic theory for SU4 breaking will be discussed taking into account such ground state impurities. This discussion will also be extended to non-doubly closed shell nuclei.

In Chap. 4, high-energy neutrino-induced reactions will be considered, and their relation to inelastic electron scattering discussed. The small cross sections of this process seem to preclude any detailed experimental study of giant resonance excitation here, but the predicted rates may serve as background estimates for a variety of elementary neutrino reactions /16-18/.

In Chap. 5, we discuss photopion reactions (γ,π) and (π,γ) on nuclei, which mainly proceed via spin-flip transitions, and hence are capable of isolating the spin-flip component of the giant resonance. Giant resonances have been observed experimentally in the radiative capture of stopped pions.

In Chap. 6, purely hadronic interactions with the nucleus will be discussed, namely (π,π'), (p,p') and (α,α'). The latter two reactions have been used in the original discovery of higher-multipole giant resonances. In these reactions, there also exists the possibility of the giant resonance entering as an intermediate state in two-step processes which manifest themselves in the backward angular distribution. When analyzed in this way /19/, additional information on the higher giant resonances may be gained. The interaction operator in the theory of hadronic interactions is not as well understood as for weak and electromagnetic interactions, so that the theoretical interpretation may be less reliable here.

In the appendices, several relevant topics involving giant resonances will be discussed such as sum rules, SU4 properties of interactions, and current algebra.

1. The Interaction Between the Nucleus and an External Probe

1.1 Nuclear Probes

The study of nuclear systems generally involves the use of external probes, i.e., the interaction with other (simpler) physical systems, the dynamics of which is better understood. A classical example of such a probe is the electromagnetic field, which intervenes, e.g., in photoreactions, or in electron scattering /13, 20, 21/. In addition, weak interactions have been successfully employed for nuclear structure studies, such as beta decay /22/ or muon capture /23/. Strongly interacting particles do not offer in general the advantages of photons, electrons, muons and neutrinos for a use as nuclear probes since they entail the problem of distinguishing between effects of nuclear structure itself and of the reaction mechanism. However, as noted by ERICSON /24/, there is an exception among these, namely, slow pions. Indeed, the scattering lengths for pion-nucleon scattering are exceptionally small: $a(\pi N) = 0.1$ fm (fm = fermi = 10^{-13} cm) in contrast to the scattering lengths for K meson- and nucleon-nucleon interactions whose values are $a(KN) = 1$ fm, $a(NN) = 10$ fm. Therefore, the interaction of slow pions with the nucleus is sufficiently weak so as not to disturb the nucleus in any violent fashion (unlike the interaction of other hadrons), which is desirable for a good probe.

In the remainder of this section, we shall consider primarily the electromagnetic and weak interactions as probes of nuclear structure, while, however, also devoting some attention to hadron interactions in the later sections of this review.

1.2 Electromagnetic and Weak Interactions of Nuclei

The usual starting point for the calculation of nuclear electromagnetic transitions /25/ is the expression for the electromagnetic nuclear 4-current density operator in the Schrödinger representation (we employ units in which $\hbar = c = 1$)

$$J^{nuc}_{em,\mu}(\underline{r}) = (\underline{J}, \rho) \tag{1a}$$

with

$$\underline{J}^{nuc}_{em}(\underline{r}) = e \sum_{j=1}^{A} \left[\frac{d}{dt} \left(\frac{1+\tau_{3j}}{2} \frac{\underline{r}_j}{2} \right) \delta(\underline{r}-\underline{r}_j) + \delta(\underline{r}-\underline{r}_j) \frac{d}{dt} \left(\frac{1+\tau_{3j}}{2} \cdot \frac{\underline{r}_j}{2} \right) \right. $$
$$\left. + \frac{1}{2m} \left(\mu_p \frac{1+\tau_{3j}}{2} + \mu_n \frac{1-\tau_{3j}}{2} \right) \delta(\underline{r}-\underline{r}_j) \, \underline{\sigma}_j \times \underline{\nabla} \right] \tag{1b}$$

(the $\underline{\nabla}$ is acting on the leptonic variables),

$$\rho^{nuc}_{em}(\underline{r}) = e \sum_{j=1}^{A} \frac{1+\tau_{3j}}{2} \delta(\underline{r}-\underline{r}_j). \tag{1c}$$

We note here that the expression of the 4-current, written above in local form, is not in general completely satisfactory, since this current is conserved only in the case of ordinary nonexchange potentials. Otherwise, it satisfies the continuity equation only in the dipole approximation, i.e.,

$$\frac{d}{dt} \int \rho(\underline{r}) \, x_k \, d^3r = - \int J_k(\underline{r}) \, d^3r. \tag{2}$$

The symbols in (1) are as follows:

e: proton charge
μ_p (μ_n): proton (neutron) magnetic moment in units of nuclear magnetons (e/2m)
$\underline{\sigma}_j$: spin operator of the jth nucleon
$\underline{\tau}_j$: isospin operator of the jth nucleon
$(d/dt)\underline{r}_j = \underline{v}_j$: velocity operator of the jth nucleon
m: nucleon mass.

The above expression for the 4-current density is approximate in the following sense:

1) the nucleons are treated as nonrelativistic,
2) they are considered to be point-like (a nucleon form factor may be inserted for high-momentum transfer reactions),
3) each nucleon is assumed to interact independently with the electromagnetic field, so that no explicit two-body currents are present, i.e., the impulse approximation is valid /26/,
4) terms of the order of q^2 have been neglected ($\underline{q}$ = momentum transfer, and $q^2 \equiv |\underline{q}|^2$).

We shall not discuss these approximations in detail here because they are standard. The matrix element between nuclear states of the Fourier transform of the electromagnetic 3-current density gives the amplitude for photon absorption by the nuclear target. In a similar way, one may describe the scattering of electrons (with initial and final wave numbers $\underline{k}_i$ and $\underline{k}_f$) from a nucleus which simultaneously undergoes a transition to an excited state /13, 20/. Indeed, one can imagine that the electron is scattered in the electromagnetic field of the nucleus A_μ^{nuc}, where $\Box A_\mu^{nuc} = J_{em,\mu}^{nuc}$.

The discussion of inelastic electron-nucleus reactions is carried out in great detail in the mentioned references. The main difference with photoreactions consists in the fact that in electron scattering, the photon absorbed by the nucleus is virtual; therefore, the transition charge density of the target also enters in the process, and $q^2 \geq \omega^2$ where q and ω are the momentum and energy of the photon, respectively. Since ω is also the nuclear excitation energy and q the momentum transfer, this means that in photoreactions where $q = \omega$, the momentum transfer is rather small ($\lesssim$ 25 MeV) but in electron scattering, it can be made much larger. It is apparent from (1) that the weight of the magnetic term grows as q increases, so that electron scattering constitutes a better means than photoreactions of observing specific effects caused by this spin-dependent operator.

We shall quote here the multipole expressions for the electromagnetic transition operators /13, 20/, and shall also specialize them to the dipole operators whose matrix elements will be of separate interest to us. The charge density may be expanded in Coulomb multipoles, while the 3-current density is expanded in transverse multipoles (both electric and magnetic). With the notations of /13, 20/ the transverse electric multipole operator is written as

$$\mathscr{T}_{LM}^{el}(q) \equiv \frac{i^L}{qe} \int d^3r \, \underline{J}_{em}^{nuc}(\underline{r}) \cdot \underline{\nabla} \times \left[j_L(qr) \, \underline{Y}_{LL}^M(\hat{r}) \right] , \tag{3a}$$

where $j_L(qr)$ is the spherical Bessel function, and $\underline{Y}_{LL}^M$, the vector spherical harmonic. The Coulomb multipole operator is

$$\mathscr{M}_{LM}^{Coul}(q) \equiv \frac{i^L}{e} \int d^3r \, \rho_{em}^{nuc}(\underline{r}) \, j_L(qr) Y_{LM}(\hat{r}). \tag{3b}$$

We note that in the long-wavelength approximation, where

$$j_L(qr) \cong (qr)^L/(2L+1)!! \tag{4}$$

the matrix elements of the isoscalar dipole operator (L = 1) vanish in the non-relativistic limit /25/ for the operator $\rho_{em}^{nuc}(\underline{r})$, see (1c). In this approximation ($q \to 0$), one has further

$$\mathscr{T}_{1M}^{el} \to \frac{\sqrt{2}}{3!!} \int d^3r \, r \, Y_{1M}(\hat{r}) \, \underline{\nabla} \cdot J_{em}^{nuc}(\underline{r}). \tag{5}$$

Using the continuity equation

$$\underline{\nabla} \cdot \underline{J} = -i\,[\mathscr{H}, \rho], \qquad (6)$$

where $\mathscr{H}$ is the nuclear Hamiltonian, we obtain the familiar result known as SIEGERT's theorem /27/, generalized to the Lth multipole

$$\langle f|\,\mathscr{T}^{el}_{LM}(q)\,|i\rangle_{q\to 0} \to \frac{\omega}{q}\left(\frac{L+1}{L}\right)^{1/2} \langle f|\,\mathscr{M}^{Coul}_{LM}(q)\,|i\rangle. \qquad (7)$$

Sometimes, the magnetic term in the current may not be negligible, e.g., for predominantly spin-flip transitions. In this case, the approximation of (7) has to be improved /28/.

Developing $\mathscr{T}^{el}_{10}(q)$ and using (1), one obtains after neglecting j_2, and approximating $j_0(qr)$ by unity and $j_1(qr)$ by $qr/3$ (unretarded dipole approximation), to first order in q^2

$$\left[\mathscr{T}^{el}_{10}(q)\right]_{q\to 0} = \frac{-1}{(4\pi)^{1/2}}\left(\frac{2}{3}\right)^{1/2} \sum_j \frac{d}{dt}\left(\frac{1+\tau_{3j}}{2}\, z_j\right)$$
$$- \frac{q^2}{(24\pi)^{1/2}}\,\frac{1}{2m} \sum_j \left(\mu_p \frac{1+\tau_{3j}}{2} + \mu_n \frac{1-\tau_{3j}}{2}\right) (\underline{r}_j \times \underline{\sigma}_j)_z. \qquad (8)$$

Strictly speaking, a correction term of order q^2 should have been retained in the first term of this equation. If used for the evaluation of isovector transitions, however, the magnetic term is enhanced by the factor $(\mu_p - \mu_n) = 4.7$, which gives justification for the form of (8).

In Chap. 2 we shall study matrix elements of the operator of (8) in inelastic electron scattering. It is worth noting that in the limit $q^2 \to 0$, the nucleons may be assumed to be point-like; however, for large values of q, the expression (1) for the 4-current should not contain the static values of the charge and the magnetic moments, but instead, the Dirac and Pauli form factors of the nucleons evaluated at the specific 4-momentum transfer of the reaction under consideration.

The transverse magnetic multipole operator is written as (/13, 20/)

$$\mathscr{T}^{mag}_{LM}(q) \equiv \frac{i^L}{e} \int d^3r\; \underline{J}^{nuc}_{em}(\underline{r}) \cdot j_L(qr)\, \underline{Y}^M_{LL}(\hat{r}). \qquad (9)$$

In the long-wavelength limit, one obtains from this for the magnetic dipole operator (L = 1)

$$\left[\mathscr{T}^{mag}_{10}(q)\right]_{q\to 0} = -\frac{1}{(6\pi)^{1/2}}\,\frac{q}{2m}\left[\sum_{j=1}^{A} \frac{1+\tau_{3j}}{2}\,\underline{l}_j + \right.$$
$$\left. + \sum_j \left(\mu_p \frac{1+\tau_{3j}}{2} + \mu_n \frac{1-\tau_{3j}}{2}\right) \underline{\sigma}_j\right] \qquad (10)$$

where $l_j = (\underline{r}_j \times \underline{v}_j) \cdot m$ is the nucleon orbital angular momentum operator.

Next, we shall consider the weak interactions of nuclei which are semileptonic interactions, described by the coupling of the weak nuclear current to the weak leptonic current of the muon or electron, and their neutrinos. Using the same approximations as for the electromagnetic current [stated after (2)], the weak nuclear current (disregarding a pseudoscalar term) is given by /29/

$$J^{nuc^{\pm}}_{weak,\mu}(\underline{r}) = J^{nuc^{\pm}}_{ax,\mu}(\underline{r}) + J^{nuc^{\pm}}_{vec,\mu}(\underline{r}) \tag{11a}$$

(Neutral currents will be introduced in Chap. 4.),

where with the convention $\tau\pm = \frac{1}{2}(\tau_1 \pm \tau_2)$

$$\underline{J}^{nuc^{\pm}}_{ax}(\underline{r}) = \sum_{j=1}^{A} \tau_j^{\pm} \; g_A \, \underline{\sigma}_j \;\; \delta(\underline{r}-\underline{r}_j) \tag{11b}$$

$$\rho^{nuc^{\pm}}_{ax}(\underline{r}) = \frac{1}{2} \sum_{j=1}^{A} \left\{ g_A \, \underline{\sigma}_j \cdot \underline{v}_j - \frac{i}{2m} g_A \, \underline{\sigma}_j \cdot \underline{\nabla}, \; \delta(\underline{r}-\underline{r}_j) \right\} \tau_j^{\pm} \tag{11c}$$

$$\underline{J}^{nuc^{\pm}}_{vec}(\underline{r}) = \sum_{j=1}^{A} \left\{ \tau_j^{\pm} \, g_V \;\; \delta(\underline{r}-\underline{r}_j) \left[\frac{\mu_p - \mu_n}{2m} \; \underline{\sigma}_j \times \underline{\nabla} \right] \right. \tag{11d}$$

$$\left. + \frac{g_V}{2} \left[\frac{d}{dt} (\tau_j^{\pm} \, \underline{r}_j) \, \delta(\underline{r}-\underline{r}_j) + \delta(\underline{r}-\underline{r}_j) \, \frac{d}{dt} (\tau_j^{\pm} \, \underline{r}_j) \right] \right\}$$

$$\rho^{nuc^{\pm}}_{vec}(\underline{r}) = \sum_{j=1}^{A} g_V \, \tau_j^{\pm} \;\; \delta(\underline{r}-\underline{r}_j). \tag{11e}$$

Here, $\underline{\nabla}$ acts on the leptonic current. One should note, however, that the approximations mentioned in obtaining (11) [stated after (2)], in particular the assumed nucleon point structure, are not strictly valid for the axial current. Its nucleon matrix element is known to have a pseudoscalar form factor g_P, essentially due to one-pion exchange with the leptonic current. (A similar term, the photoelectric term, appears in pion photoproduction, to be discussed below). The pseudoscalar term has a strongly momentum-transfer dependent matrix element; and the finite nucleon size even causes momentum-transfer dependent form factors to accompany g_V and g_A which may, however, be taken constant for moderately low values of $q \lesssim 100$ MeV/c (in contrast to g_P). In (11), the assumptions of the V-A interaction, the conserved vector current hypothesis, and muon-electron universality have been used (see, e.g. /22/).

In the weak interaction, a multipole expansion may be performed also. In Chaps. 3 and 4, we shall be mainly interested in the Fermi and Gamow-Teller dipole operators, given by the expressions $\sum \tau_j^{\pm} \, \underline{r}_j$ and $\sum_j (\underline{\sigma}_j \otimes \underline{r}_j) \, \tau_j^{\pm}$, respectively.

In the expansion in powers of q of the Fourier transform of the weak current, the dipole terms are those which contain the first order in q. Therefore, they contribute to 1st forbidden transitions, while the zeroth-order term is called the allowed contribution.

1.3 Excitation Operators and Nuclear Structure

The experimental study of nuclear structure with electromagnetic and weak probes consists essentially of two kinds of experiments:

1) Elastic scattering, in which the final nuclear state is again the ground state.
2) Inelastic scattering (or photoabsorption processes) where the nuclear final state is an excited state.

Among other processes, we shall study the weak and electromagnetic excitation of doubly-closed-shell nuclei ^4He, ^{16}O and ^{40}Ca to T = 1 negative parity levels. From experimental data /30/ the squares of the allowed weak matrix elements for these nuclei turn out to be ~1/100 of the squared weak matrix elements of other nuclei in general, therefore the 1st forbidden weak operators play a dominant role /31/. All these operators, i.e., $\sum_j \tau_j^{\pm} \underline{r}_j$, $\sum_j \tau_j^{\pm} (\underline{r}_j \otimes \underline{\sigma}_j)$ together with their electromagnetic analogs $\sum_j \tau_{3j} \underline{r}_j$ and $\sum_j \tau_{3j} (\underline{r}_j \otimes \underline{\sigma}_j)$ which are dipole operators and dipole-spin-flip operators, respectively, give rise to giant resonance phenomena, i.e., to large peaks in the differential cross section $d\sigma(\omega)/d\omega$ (where ω: energy loss of the probe = nuclear excitation energy) at excitation energies ≈20 MeV /32/. The corresponding excited states (giant resonance states: giant resonance and spin-flip giant resonance) are usefully classified in the framework of the SU4 approximate symmetry, meaning approximate spin and isospin independence of the nuclear Hamiltonian /33/. This classification is possible due to the fact that:

1) these operators have a definite tensor character (vector) under this group, the generators of which are

$$T_\alpha = \frac{1}{2}\sum_{j=1}^{A} \tau_{\alpha j}$$

$$S_\lambda = \frac{1}{2}\sum_{j=1}^{A} \sigma_{\lambda j} \tag{12}$$

$$Y_{\alpha\lambda} = \frac{1}{2}\sum_{j=1}^{A} \tau_{\alpha j}\,\sigma_{\lambda j}$$

(the operators commute with a spin and isospin independent Hamiltonian) /34/,

2) the ground states of doubly closed shell nuclei are approximately scalar supermultiplets as a consequence of the short-range attractive character of nuclear

forces which favors the maximum spatial symmetry and hence, due to the Pauli principle, the spin and isospin antisymmetry. As a consequence, calculated values of the allowed transitions in beta decays are small for doubly closed shell nuclei /35/, in agreement with experiment. Indeed, in the SU4 limit they would be completely forbidden because

$$\sum_j \tau_j |0\rangle = 0 \, , \quad \sum_j \sigma_j \tau_j |0\rangle = 0 .$$

if $|0\rangle$ is a scalar supermultiplet. Their actual values, however, depend on the amount of SU4 breaking /36, 37/. From beta decays, one obtains an order of magnitude estimate of ~1% for the squared admixture of nonscalar impurities in supermultiplets /30/. Also, electromagnetic M1 transitions may be a useful tool for the estimate of this admixture. NANG /38/ indeed obtained for the squared impurity of the ground states the values ~20% for ^{12}C, <3% (^{16}O), 1% (^{4}He and ^{40}Ca). The theory of reactions involving ^{12}C will therefore require a more careful treatment (see Chap. 2).

Experimentally, giant resonance phenomena have been observed in photoreactions /2, 3/, in electron scattering /13, 21/, muon capture /23, 34/, and radiative pion capture /39 - 41/. The nuclear photoabsorption cross section (see Fig. 1) is

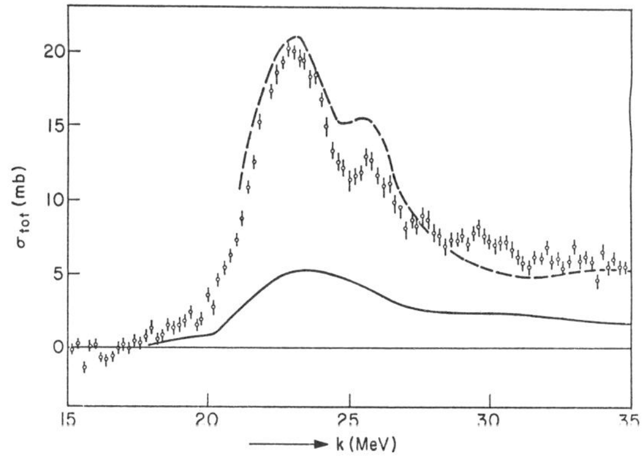

Fig. 1.(a) Total photoabsorption cross section in ^{12}C as measured by the Mainz group (J. AHRENS et al., International Photonuclear Conference, Asilomar, March 1973), showing theoretical dipole (dashed curve) and higher multipole (and retarded dipole) (solid curve) estimated contribution /12/

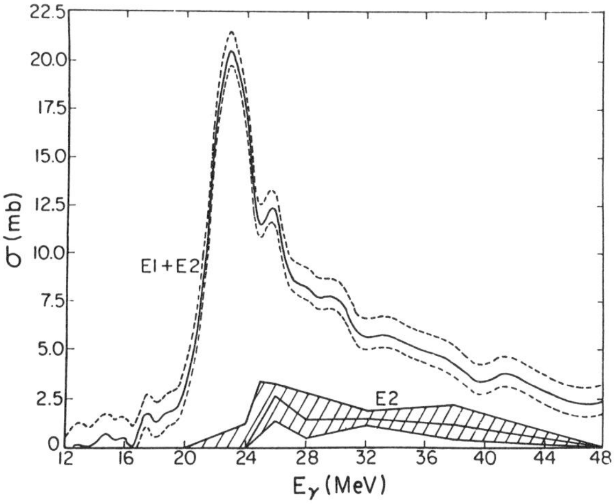

Fig. 1.(b) Experimental (E2) background /43/

characterized by a peak having a width of about 5 MeV, with mean energy which decreases from ~22 MeV to ~13 MeV going from light to heavy nuclei. The theoretical interpretation that the main photoabsorption peak is given by the dipole excitation of the nucleus is supported by the fact that the photoabsorption cross section, integrated over this peak, exhausts a large fraction (generally ≳70 %) of the value expected from the dipole sum rule /44/ in absence of exchange and velocity dependent nucleon-nucleon potentials,

$$\int_0^\infty \frac{d\sigma(\omega)}{d\omega}\, d\omega = \frac{2\pi^2 e^2}{m}\,\frac{NZ}{A} = \frac{60\ NZ}{A}\ [\mathrm{MeV}\cdot\mathrm{mb}], \tag{13}$$

where N and Z are the neutron and proton number, respectively, of the nucleus, and $A = N + Z$. The observed peak, caused by excitation via the isospin dipole operator $\sum_j \tau_{3j}\, \underline{r}_j$, was referred to as the giant resonance of the photonuclear effect.

With electron-nucleus scattering, further details of the giant resonance may be explored. One may here measure the two-dimensional surfaces $d\sigma/d\omega dq^2$ over the momentum transfer and energy loss plane and, integrating over the giant resonance energy region, one observes both isospin and spin-isospin states [excited by the operators $\sum_j \tau_{3j}\, \underline{r}_j$ and $\sum_j \tau_{3j}\, (\underline{r}_j \times \underline{\sigma}_j)$, respectively] which vary in their relative strength due to a different q dependence of the isospin excitation and the spin-isospin excitation /13, 20, 28/.

The total isospin is a good quantum number in nuclear physics /45/; therefore, one can very simply relate charge conserving matrix elements (as in electromagnetic interactions) to the charge changing ones (as in weak interactions). In this manner one can reduce the operator describing the 1st forbidden part of the vector weak interaction $\sum_j \tau_j^{\pm}\, \underline{r}_j$ to $\sum_j \tau_{3j}\, \underline{r}_j$. The latter is exactly the operator describing photonuclear absorption (as follows from Siegert's theorem and from the fact that in the nonrelativistic limit the isoscalar dipole matrix elements between nuclear states vanish in the long-wave approximation). These two operators (and hence, their matrix elements) can be connected through the commutator with the (∓) operator of the total isotopic spin of (12) as follows

$$\left[\sum_j \tau_{3j}\, \underline{r}_j,\ T^{\mp}\right] = \mp\, 2 \sum_j \tau_j^{\mp}\, \underline{r}_j \qquad (T^{\pm} = T_1 \pm i\, T_2). \tag{14}$$

Since the dominant feature of photoabsorption is the giant dipole resonance, we expect a similar phenomenon in muon capture on doubly closed shell nuclei /23, 34/ at least as far as the weak vector current matrix elements are concerned. These arguments may be extended to the axial dipole matrix elements (Chap. 3). The pioneering work in this field is due to FOLDY and WALECKA /31/. These authors related muon capture rates to the experimentally measured photonuclear cross section, taking therefore automatically into account the giant resonance excitation. For such a procedure, one assumes that the nuclear wave functions belong to a particular irreducible representation of some appropriate group (a symmetry group or at least an approximate symmetry group) under which the nuclear transition operators of interest can also be classified. In particular, the group SU4 has been used to study the relations between electron scattering, beta decay, photoabsorption and muon capture. The giant resonance states are assumed to belong to a vector supermultiplet generated by the application of the isovector dipole operator (with or without spin flip) to the scalar supermultiplet (ground state).

The giant resonance excitation can be pictured as dipole nuclear matter vibrations. They are either of isospin character (i.e., excited by the operator $\sum_j \tau_{3j} \underline{r}_j$) and appear strongly in photoabsorption or low-q electron scattering, or they are of spin-isospin type (i.e., excited via $\sum_j \tau_{3j} \underline{r}_j \otimes \underline{\sigma}_j$) and are observed strongly at higher q values /21/. The isospin vibrations are excited also by the Fermi transitions of weak interactions, and the spin-isospin vibrations through the Gamow-Teller transitions. The energies of all these vibrations are degenerate in the SU4 limit /32/. A model based on these dipole vibrations is the so-called generalized Goldhaber-Teller model /46/ (the original Goldhaber-Teller model had been confined to electric dipole vibrations of neutrons against protons only /47/). This model will be discussed in Sect. 1.4 since it permits a useful SU4 classification of the giant resonances; in addition, we shall there describe its extension to giant multipoles, and their interpretation in terms of collective surface waves.

Besides by the Goldhaber-Teller model, all these giant resonance phenomena, with their observed fine structure, may be explained in principle by detailed shell model calculations with residual interactions using a sufficiently large configuration space (see /48/ for photoabsorption, /49/ for electron scattering, /23, 34/ for muon capture, /41/ for radiative pion capture).

In general, when calculating rates for the processes mentioned above, one deals with expressions of the type

$$\sum_n |\langle n| \sum_j \tau_j \underline{r}_j |0\rangle|^2 f(n), \tag{15}$$

i.e., summed over the excited states $|n\rangle$ with energy E_n, and where the functions f(n) take into account phase space, etc. The sum over $|n\rangle$ runs over all energetically accessible states of the daughter nucleus only, but this energy constraint is not strong in many physical cases, e.g., muon capture and radiative pion capture. Energetically accessible states are very numerous and in addition, the absolute values of the matrix elements $|\langle n| \sum_j \tau_j \underline{r}_j |0\rangle|^2$ are largest for the states of the giant dipole supermultiplet, lying at about 20 MeV for the doubly closed shell nuclei (which is small compared to the ~100 MeV available in the mentioned reactions). In this situation, one may change the sum over all energetically accessible states into an unrestricted sum over all states, a procedure known as the closure approximation /34/. The use of sum rules /50/ then relates the expression of (15) to a ground state expectation value. Using these techniques, one may evaluate, e.g.,

$$\sum_n \omega_n^q \, |\langle n|\mathcal{O}|0\rangle|^2 = \langle 0|\mathcal{O}^+ \mathcal{H}^q \mathcal{O}|0\rangle, \tag{16}$$

where $\mathcal{O}$ is the transition operator, $\omega_n = E_n - E_0$ is the excitation energy, $\mathcal{H}$ the Hamiltonian of the target nucleus, and f(n) must be developed in powers of ω_n.

When a transition shows a giant resonance excitation, sum rules are very useful, because they may be largely saturated already with the giant resonance peak. Furthermore, the presence of giant resonance analogs (also those of spin-isospin type) is predicted by supermultiplet theory so that this scheme implements very well sum rule techniques /32/. For the dipole operators $\sum_j \tau_j \underline{r}_j$, $\sum_j (\underline{\sigma}_j \otimes \underline{r}_j)\, \tau_j$ the sum rule approach, if saturation and degeneracy between all giant resonance components are assumed, gives the same results as the generalized Goldhaber-Teller model /46/.

1.4 Classification of Giant Multipoles and Nuclear Models

The classification of nuclear giant dipole resonances can be carried out by considering the SU4 group structure of the excitation operators (12). It may alternately be obtained from the use of a suitable nuclear model, in particular the generalized Goldhaber-Teller model /46/, which will be briefly discussed here. In its original form /47, 51/ this model, in order to obtain the observed coherent collective transition dipole moment of the nucleus in photoexcitation processes, assumed a harmonic vibration of the protons as a whole against the neutrons as a whole (i oscillations), with a displacement vector $\underline{d}$. A rigid displacement of the charge density $\rho_0(r)$ leads to

$$\rho_0 \left(|\underline{r} - \tfrac{1}{2}\underline{d}|\right) \cong \rho_0(r) - \tfrac{1}{2}\underline{d} \cdot \underline{\nabla}\, \rho_0(r) \tag{17}$$

(in the approximation $Z \cong N$). With $\underline{d}$ being considered the coordinate of a harmonic oscillator with a Hamiltonian

$$H = (1/2\, m^*)\, \underline{p}^2 + \tfrac{1}{2} m^* \omega^2 \underline{d}^2 \tag{18}$$

(reduced mass $m^* = A\, m/4$, experimental eigenfrequency ω), it may upon quantization of the latter be expressed in the form (here L = 1)

$$d_{Lm} = (1/2\, m^* \omega_L)^{1/2} \left[a^+_{L,m} + (-)^m a_{L,-m}\right], \tag{19}$$

i.e., in terms of creation (a^+_{Lm}) and annihilation (a_{Lm}) operators acting on states that span the space of the dipole phonons.

There are, however, other collective modes of oscillation, degenerate with the Goldhaber-Teller modes to the extent that the nuclear forces are approximate spin and isospin independent (so that WIGNER's supermultiplet theory is applicable /33/) that may be described as the oscillation of a sphere of protons with spin in a given direction and neutrons with spin in the opposite direction, against a sphere of protons with spin in the opposite direction, and neutrons with spin in the given direction (spin-isospin or "si-oscillations"), and furthermore oscillations of a sphere of protons and neutrons with spin in a given direction against a sphere of protons and neutrons with spin in the opposite direction (spin waves, s). These latter two types of collective vibrations were first obtained by WILD /52/, and later independently by GLASSGOLD et al. /53/.

The nuclear transition density matrix (in spin-isospin space) $\phi(\underline{r})$ of the three modes of oscillation discussed above is accordingly:

$$\phi_i(\underline{r}) = \tfrac{1}{2}(1+\tau_3)\,\tfrac{1}{2}\,\rho_o(\underline{r}-\tfrac{1}{2}\,\underline{d}) + \tfrac{1}{2}(1-\tau_3)\,\tfrac{1}{2}\,\rho_o(\underline{r}+\tfrac{1}{2}\,\underline{d}) \tag{20a}$$

$$\begin{aligned}\phi^{m'}_{si}(\underline{r}) = {} & \tfrac{1}{4}\left[(1+\tau_3)(1+\sigma_{m'}) + (1-\tau_3)(1-\sigma_{m'})\right]\tfrac{1}{2}\,\rho_o(\underline{r}-\tfrac{1}{2}\,\underline{d}) \\ & + \tfrac{1}{4}\left[(1+\tau_3)(1-\sigma_{m'}) + (1-\tau_3)(1+\sigma_{m'})\right]\tfrac{1}{2}\,\rho_o(\underline{r}+\tfrac{1}{2}\,\underline{d})\end{aligned} \tag{20b}$$

$$\phi^{m'}_{s}(\underline{r}) = \tfrac{1}{2}(1+\sigma_{m'})\,\tfrac{1}{2}\,\rho_o(\underline{r}-\tfrac{1}{2}\,\underline{d}) + \tfrac{1}{2}(1-\sigma_{m'})\,\tfrac{1}{2}\,\rho_o(\underline{r}+\tfrac{1}{2}\,\underline{d}). \tag{20c}$$

The values m' = 1,0, -1 correspond to the triplet states of the final total spin $\underline{S}$ which is uncoupled from the spatial motion in our model, and in the supermultiplet theory in general. Expanding and keeping terms of first order in $\underline{d}$ only, which describe the $0^+ \to 1^-$ transition to the oscillator dipole states, we find

$$\phi_i(\underline{r}) = -\,\tfrac{1}{4}\,\tau_3\,\underline{d}\cdot\underline{\nabla}\,\rho_o(\underline{r}) \tag{21a}$$

$$\phi^{m'}_{si}(\underline{r}) = -\,\tfrac{1}{4}\,\tau_3\sigma_{m'}\,\underline{d}\cdot\underline{\nabla}\,\rho_o(\underline{r}) \tag{21b}$$

$$\phi^{m'}_{s}(\underline{r}) = -\,\tfrac{1}{4}\,\sigma_{m'}\,\underline{d}\cdot\underline{\nabla}\,\rho_o(\underline{r}). \tag{21c}$$

The transition densities of isospin, spin-isospin and spin being

$$\rho_i(\underline{r}) = \tfrac{1}{2}\,\langle f|\sum_{i=1}^{A}\delta(\underline{r}-\underline{r}_i)\tau_{3i}|0\rangle \tag{22a}$$

$$\underline{\rho}_{si}(\underline{r}) = \tfrac{1}{2}\,\langle f|\sum_{i=1}^{A}\delta(\underline{r}-\underline{r}_i)\,\underline{\sigma}_i\,\tau_{3i}|0\rangle \tag{22b}$$

$$\varrho_s(\underline{r}) = \frac{1}{2} <f| \sum_{i=1}^{A} \delta(\underline{r}-\underline{r}_i)\ \underline{\sigma}_i |0>, \tag{22c}$$

respectively, are obtained from the density matrix $\phi(\underline{r})$ by

$$\rho_i(\underline{r}) = \frac{1}{2} \mathrm{Tr}\ \ \tau_3\ \phi_i(\underline{r}) \tag{23a}$$

$$\varrho_{si}(\underline{r}) = \frac{1}{2} \mathrm{Tr}\, \underline{\sigma}\ \tau_3\ \phi_{si}^{m'}(\underline{r}) \tag{23b}$$

$$\varrho_s = \frac{1}{2} \mathrm{Tr}\, \underline{\sigma}\ \phi_s^{m'}(\underline{r}) \tag{23c}$$

and we find using (22)

$$\rho_i(\underline{r}) = - \frac{1}{2} \underline{d} \cdot \underline{\nabla}\ \rho_0(r), \tag{24a}$$

i.e., the transition part of (17), which gets its contribution from the isospin mode only, and

$$\rho_{si}^{j}(\underline{r}) = \rho_s^{j}(\underline{r}) = - \frac{1}{2} \delta_{jm'}^{(S)}\ \underline{d} \cdot \underline{\nabla}\ \rho_0(\underline{r}) \tag{24b}$$

using the spherical Kronecker tensor

$$\delta_{jm'}^{(S)} = \begin{cases} -\ 2^{-1/2}\ (\delta_{j1} + i\delta_{j2}) & m' = 1 \\ \delta_{j3} & m' = 0 \\ 2^{-1/2}\ (\delta_{j1} - i\delta_{j2}) & m' = -1 \end{cases} \tag{24c}$$

which arise only from the si mode or the s mode, respectively. The correspondence of (21) with the SU4 generators of (12) is evident, but this structure has now been derived from an explicit nuclear model, i.e., the generalized Goldhaber-Teller model for the dipole vibrations of nuclear matter. Note that while $\rho_i(\underline{r})$ describes transitions $0^+ \rightarrow 1^-$ of even nuclei, ϱ_{si} and ϱ_s will lead to transitions $0^+ \rightarrow J^-$ of even-even nuclei, where the total spin $\underline{S} = 1$ couples with the dipole matter oscillation of angular momentum $\underline{L} = 1$ to the final nuclear spins $J^\pi = 0^-$, 1^-, and 2^-.

This model may readily be generalized /12/ to collective nuclear-matter oscillations of higher multipolarity ($L > 1$). Here, the ground state density $\rho_0(\underline{r})$ is assumed to be deformed by a scale factor,

$$\rho_0\ r(1-\eta)\ \ \cong \rho_0(r) - \eta r\ [d\rho_0(r)/dr]^{(L)}, \tag{25a}$$

where we use a derivative

$$[d\rho_0/dr]^{(L)} = r^{-3}(d/dr)(r^3\rho_0), \qquad L = 0$$
$$= d\rho_0/dr \qquad L \geq 1 \tag{25b}$$

defined in such a way as to produce matter conservation for the case of a monopole, L = 0 /13/. The scale factor η is expanded in a series of multipoles,

$$\eta = \sum_{Lm} \alpha_{Lm} (r/R)^{L+2\delta_{L0}-2} Y^*_{Lm}(\hat{r}), \tag{25c}$$

R being a reference radius which renders η dimensionless. The choice of the power of r in (25c) corresponds to the assumption of an incompressible (except for L = 0), irrotational fluid flow /13/. In analogy with (21), we now obtain transition density matrices

$$\phi_i(\underline{r}) = -\frac{1}{2}\tau_3\,\eta r\,[d\rho_0(r)/dr]^{(L)} \tag{26a}$$

$$\phi^{m'}_{si}(\underline{r}) = -\frac{1}{2}\tau_3\,\sigma_{m'}\,\eta r\,[d\rho_0(r)/dr]^{(L)} \tag{26b}$$

$$\phi^{m'}_{s}(\underline{r}) = -\frac{1}{2}\sigma_{m'}\,\eta r\,[d\rho_0(r)/dr]^{(L)} \tag{26c}$$

$$\phi_c(\underline{r}) = -\frac{1}{2}\eta r\,[d\rho_0(r)/dr]^{(L)} \qquad (L \neq 1) \tag{26d}$$

where a fourth mode (compressional) appears which describes nuclear matter motion of all four nucleon fluids in unison. It is seen that nuclear matter is here described in terms of four interpenetrating fluids, $p^{\uparrow}$, $p_{\downarrow}$, $n^{\uparrow}$ and $n_{\downarrow}$ (in a self-evident notation). As for the dipole case, the SU4 structure of these density N form matrices is obvious: ϕ_i, ϕ_{si} and ϕ_s form a 15-dimensional SU4 vector operator while ϕ_c is an SU4 scalar. (Note that "density vibration" would be a better designation for this mode).

In Table 1, we present a list of the possible states in a giant resonance supermultiplet for a self-conjugate nucleus ($J^{\pi}_{g.s.} = 0^+$) corresponding to the 2^L pole vibration of nuclear matter. The combinations of S and T values of the transition lead to the possible final nuclear spins J^{π} as shown (with states of different J^{π} possibly split by spin-orbit coupling /54/), and to the number of states N in each mode which form the components of a 15-component SU4 vector or of an SU4 scalar as indicated.

Table 1. The giant resonance SU4 supermultiplet of a self-conjugate nucleus which undergoes 2^L pole matter vibrations, N = multiplicity

	Mode	T	S	J^π	N	Total
$L^{\pi=(-)^L} > 0$	i	1	0	L^π	3	
	si	1	1	$(L-1)^\pi$, L^π, $(L+1)^\pi$	9	15
	s	0	1	$(L-1)^\pi$, L^π, $(L+1)^\pi$	3	
	c	0	0	L^π	1	1
$L^\pi = 0^+$	i	1	0	0^+	3	
	si	1	1	1^+	9	15
	s	0	1	1^+	3	
	c	0	0	0^+	1	1

The components of the giant resonance SU4 supermultiplet may be compared to the resonances which are observed experimentally in various intermediate-energy reactions /55/. It is obvious that different components will be excited in a different fashion by different types of reactions, depending upon which of the SU4 generators of (12) enters in the theoretical transition operators that describe that particular reaction. For example, photonuclear reactions excite predominantly the i mode; radiative pion capture and photopion reactions with charged pions mainly the spin-isospin (si) mode; electron scattering, muon capture and neutrino excitations mainly the i and si modes (T = 1). Alpha particle scattering leads to T = 0 modes while proton and pion scattering may excite all modes but with different weights depending on the scattering angle. The feature of the various intermediate-energy processes to select certain substates of the supermultiplet may be considered a decisive advantage since in this fashion, the SU4 character of a given resonance may be better ascertained depending on whether or not it is excited in any of the mentioned reactions. For this purpose, reactions in which the momentum transfer $|\underline{q}|$ may be varied (e.g., in electron, pion, etc., scattering) are superior to those of fixed momentum transfer (photonuclear reactions with $|\underline{q}| = \omega$, muon capture with $|\underline{q}| \cong m_\mu$, radiative capture of stopped pions with $|\underline{q}| \cong m_\pi$) because the former can obtain an entire form factor as a function of $|\underline{q}|$ which will provide further information (e.g., on the multipolarity L^π) which will identify the character of the observed resonance.

Other giant resonance models of collective type (as contrasted with shell-model type models which usually incorporate spin-orbit coupling, so that the SU4 character of the final states generated is somewhat obscured /13, 46/) represent alternatives to, or refinements of the Goldhaber-Teller model and could equally well be employed for SU4 classification purposes. Curiously enough, this point seems not to have been considered in the literature as yet, so that we shall restrict ourselves here only to a brief discussion of these models.

The STEINWEDEL-JENSEN model /56, 57/ as formulated for the isospin mode (the other SU4 modes have not been discussed but could easily be formulated also) assumes two interpenetrating fluids of neutrons and protons with constant total density,

$$\rho_p + \rho_n = \rho_o \qquad (r \leq R_o) \tag{27a}$$

enclosed in a rigid spherical surface of radius $R_o = r_o A^{1/3}$. Collective motion is described by an excess density $\eta(\underline{r}, t)$

$$\begin{aligned} \rho_p(\underline{r}, t) &= (Z/A)\rho_o + \eta(\underline{r}, t) \\ \rho_n(\underline{r}, t) &= (N/A)\rho_o - \eta(\underline{r}, t). \end{aligned} \tag{27b}$$

The potential energy of this motion is provided by the symmetry energy

$$E_{sym} = K(N-Z)^2/A \tag{28}$$

($K \cong 25$ MeV) from the semiempirical mass formula. Variation of the corresponding Lagrangian leads to the wave equation

$$(\nabla^2 + k^2)\eta(\underline{r}) = 0, \tag{29}$$

where a time dependence $\exp(i\omega t)$ is used. The sound velocity in nuclear matter is obtained as

$$u = \omega/k = (8KZN/A^2 m)^{1/2}. \tag{30}$$

With the rigid-surface boundary condition

$$[\partial\eta/\partial r]_{r=R_o} = 0 \tag{31}$$

(no outflow at $r = R_o$) one finds the eigenmodes

$$\eta_{lmn}(\underline{r}) = A_{lmn}\, j_l(K_{ln} r) Y_{lm}(\hat{r}) \tag{32a}$$

$$A_{lmn} = \left\{ j_l(K_{ln} R_o) \left[\frac{1}{2} R_o^3 \left(1 - \frac{l(l+1)}{K_{ln}^2 R_o^2} \right) \right]^{1/2} \right\}^{-1}, \tag{32b}$$

where the eigenfrequencies are determined by (31), i.e.,

$$\left[(d/dr)j_1(K_{1n}r)\right]_{r=R_0} = 0 \qquad (32c)$$

(n labeling the roots). This leads, from the first node $K_{10}R_0 = 2.08$, to the giant dipole energy

$$\omega_{10} = (2.08/R_0)(8KNZ/mA^2)^{1/2} \sim 80\ A^{-1/3}\ \text{MeV}. \qquad (33)$$

The model may be quantized as for the case of the Goldhaber-Teller model. It may be shown /13/ that the Steinwedel-Jensen model exhausts 87% of the photonuclear sum rule, while the Goldhaber-Teller model exhausts 100% /32, 46/. It is clear that instead of the two fluids (p, n) considered here, the four fluids $p^{\uparrow}$, $p_{\downarrow}$, $n^{\uparrow}$, $n_{\downarrow}$ used in the Goldhaber-Teller model could equally well have been introduced here, leading to an equivalent SU4 classification in the Steinwedel-Jensen model.

The two models can be combined /58/ starting from the following considerations. In a Steinwedel-Jensen-type density vibration, the inertia associated with the flow of neutrons and protons would tend to carry them beyond the location of the original surface when they get displaced first to one side of the nucleus and then to the other. This tendency of the neutron and proton boundaries to undergo a harmonic displacement against each other just represents the Goldhaber-Teller model. The displacement beyond the nuclear boundary cannot be avoided except by the unrealistic assumption of an infinitely stiff restoring force resisting such displacements. It has been found /58/ that a good description of the giant resonance may be obtained by admixing to the Goldhaber-Teller mode a Steinwedel-Jensen mode whose weight increases for heavier nuclei. It should be noted that in order to have a realistic description of the transition density, the simple-minded collective predictions should be improved because one has to take into account the fact that the transition strength is actually fragmented due to nucleon-nucleon residual interactions, which may influence in a different way the various spin-isospin transition modes.

For N > Z nuclei, there arises the question what role will be played by the neutron excess fluid. Disregarding spin effects, this topic has been treated on the basis of a "three-fluid model" by MOHAN et al. /59/. The three fluids are the protons, the neutrons in the same orbitals as the protons (blocked neutrons), and the excess neutrons, to account for the fact that the excess neutrons interact in a different way with protons. This automatically takes into account effects such as Pauli blocking, etc.

The three-fluid model generalizes the model of Steinwedel and Jensen. The nuclear fluid is assumed to be incompressible, and to be distributed inside a sphere of radius $R_0 = r_0\ A^{1/3}$; the three fluids by themselves are compressible. While in the two-fluid model, the neutrons have to move in a direction opposite to that of protons to preserve the nuclear matter density, in the three-fluid model the protons and the blocked neutrons can both move toward the surface without violating the incompressibility because the excess neutrons, i.e., the third fluid, can move in toward the nuclear center to take their place /59 - 61/. This additional mode is related to the more complex isospin structure of the dipole vibrations for $N > Z$ nuclei.

To complete these general remarks, we note that due to the isovector nature of the E1 photoexcitation operator, the giant dipole resonances seen in self-conjugate (ground state $T_0 = 0$) nuclei are $T = 1$ states, while in $T_0 \neq 0$ nuclei they consist of separate $T_> = T_0 + 1$ and $T_< = T_0$ components expected to be split by several MeV (with $T_>$ lying higher, and being an analog state, which has a counterpart in the neighboring nucleus with ground state isospin $T_0 + 1$ /62/). Particularly interesting are here photonuclear reactions in light nuclei with one or two nucleons outside a closed shell, since experimental results /63/ can be compared in this case with microscopic shell-model predictions /64/. If there is no isospin mixing, then starting from a closed-shell nucleus ($T_0 = 0$) and a valence neutron ($T = 1/2$) such as ^{17}O, one can excite by photoreactions both $T = 3/2$ and $T = 1/2$ states. However, the decay $T = 3/2 \rightarrow$ (closed-shell nucleus)$_{T_0=0}$ plus neutron is forbidden, so that one can here study $T = 1/2$ strength by observing (γ, n_0) spectra. The results of a shell-model calculation /64/ for the photoexcitation of the $T = 1/2$ and $T = 3/2$ states in ^{17}O is shown in Fig. 2, together with experimental ^{17}O (γ, n_{total}) cross sections /63/.

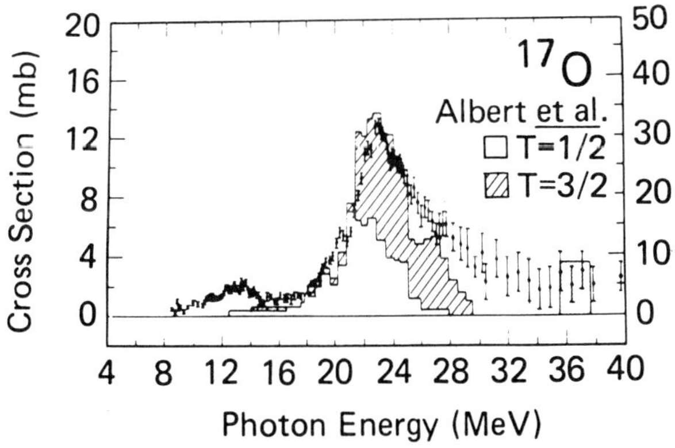

Fig. 2. ^{17}O (γ, n_{tot}) cross section /63/ compared with the theoretical result of /64/ calculated using a δ potential with Soper exchange mixture. The calculated result is the total photoabsorption cross section (right-hand scale in mb).

2. Dipole and Multipole Giant Resonances in Electron Scattering

2.1 Kinematics and Cross Section

As a probe of nuclear structure, electron scattering experiments have reached a high degree of energy resolution, so that they are particularly suited for fine structure investigation (e.g., for determination of energy and width of the excited states, their quantum numbers, etc.). They provide knowledge of the multipolarity and the charge (longitudinal or Coulomb) and 3-current (transverse) matrix elements or "form factors" of a given level.

In order to study nuclear transitions of definite multipolarity, the following one-photon exchange (Born approximation) expression of the cross section is useful /13, 20/:

$$\frac{d\sigma}{d\Omega} = 4\pi\sigma_M \left(\frac{\Delta^2}{q^2}\right)^2 \left\{F_C^2(q) + \frac{q^2}{\Delta^2}\left[\frac{1}{2} + \frac{q^2}{\Delta^2}\tan^2\frac{\theta}{2}\right] F_T^2(q)\right\} , \tag{34a}$$

the Mott cross section being

$$\sigma_M = \left(\frac{\alpha}{2k_i}\right)^2 \left(\frac{\cos\theta/2}{\sin^2\theta/2}\right)^2 , \tag{34b}$$

where the electron energy is assumed large compared to its mass, $\alpha = 1/137$, and the momentum transfer $\underline{q} = \underline{k}_i - \underline{k}_f$. Here, $\underline{k}_{i,f}$ are the initial and final electron momentum, $\theta = \sphericalangle(\underline{k}_i, \underline{k}_f)$ is the scattering angle, and $\Delta^2 = q^2 - \omega^2$ where ω is the energy of the electroexcited level.

The quantities

$$F_C^2(q) = \hat{J}_0^{-2} \sum_{L=0}^{\infty} |\mathscr{M}_L^{\mathrm{Coul}}(q)|^2 \tag{35a}$$

and

$$F_T^2(q) = \hat{J}_0^{-2} \sum_{L=1}^{\infty} \left\{|\mathscr{T}_L^{\mathrm{el}}(q)|^2 + |\mathscr{T}_L^{\mathrm{mag}}(q)|^2\right\} \tag{35b}$$

are called the Coulomb and transverse nuclear form factors, respectively, where $\hat{J}_0^2 = 2J_0 + 1$, J_0 being the initial nuclear spin. The quantities $\mathcal{M}_L^{Coul}$ and $\mathcal{T}_L^{el,mag}$ in this formula are here understood to be the angular-momentum-reduced nuclear matrix elements of the corresponding operators defined in (3b,a and 9), respectively, taken between the initial state $|J_0M_0\rangle$ (ground state) and a definite final state $|J_fM_f\rangle$.

The kinematical factors in (34a) are such that for backward electron scattering ($\theta \to 180°$), only the transverse matrix elements contribute to electron scattering.

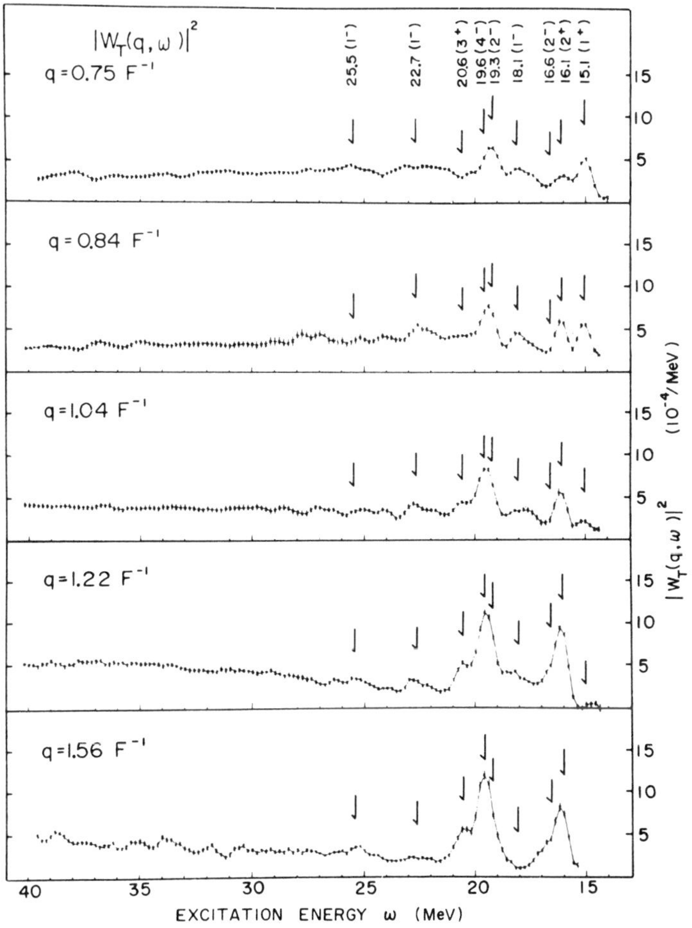

Fig. 3. Squared transverse electron scattering form factor of ^{12}C vs excitation energy, measured at various constant values of q as indicated /65/

2.2 Giant Dipole Excitation in Even-Even Nuclei

Experimentally, electroexcitation cross section of (34) may be measured at a fixed incident electron energy E_i as a function of scattering angle θ, or at a fixed angle as a function of energy. It is more advantageous, however, to perform measurements while varying both E_i and θ in such a way that the momentum transfer

$$q \cong 2k_i \sin\frac{\theta}{2}$$

(for $k_i >> \omega$) is kept constant, because then the form factors of (35) do not vary and the longitudinal and transverse terms in (34) may be determined separately /65/. If, e.g., the transverse cross section is measured in that way, one obtains the transverse excitation spectrum of the nucleus as a function of excitation energy $\omega = k_i - k_f$ as shown in Figure 3. The transverse giant resonance form factor

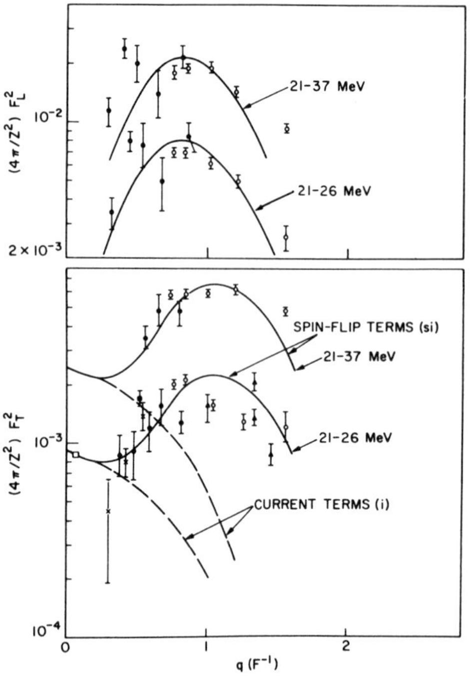

Fig. 4. Squared longitudinal (top) and transverse (bottom) form factor of the giant resonance of ^{12}C vs q, integrated over two different regions of excitation energy as indicated, and fit by the Helm model /66/

(integrating the data of Figure 3 from ω = 21 MeV upwards) is shown in Figure 4 plotted vs q in the lower portion (and the longitudinal form factor in the upper portion) for upper limit of integration ω = 26 or 37 MeV, together with curves giving a phenomenological interpretation by means of the Helm model /66/, as will be discussed in Sect. 4.5.

The cross section of (34) represents the total strength of the transition $|J_0> \to |J_f>$, integration over the width of the final state being implied. If the final state is expected to largely saturate the strength of a particular multipole operator causing the transition, say $\mathscr{T}_1^{el}$ for the case of the electric dipole giant resonance, then (35b) may be considered a transverse "sum rule" /28/ for this particular multipole transition. Although this in principle would include elastic scattering, that contribution to the sum rule vanished for nuclei with ground-state spin J_0 = 0 which have a vanishing ground state magnetic moment /13/.

For various light even-even nuclei, there exists experimental evidence of a dip in the integrated giant resonance form factor at q ~ 60 MeV/c when plotted vs q. This dip is pronounced for nuclei with closed subshells only (e.g., ^{12}C /65, 67/, see Figure 4; or ^{28}Si /68/), but it is quite shallow for double closed shell nuclei (e.g., ^{16}O /69, 70/). The existence of these dips is theoretically well established by shell-model calculations /71, 72/, but it has also been shown using closure approximation techniques that the mentioned differences in the dip depth for different nuclei may be explained /73/ by the various ground state properties of these nuclei, cf. (16). In fact, the depth of the dip depends on the ground-state expectation value of the operator $\sum_j \underline{l}_j \cdot \underline{s}_j$, ($\underline{s}_j = 1/2\, \underline{\sigma}_j$) thus providing in principle a means of determining this quantity which also enters in the Kurath sum rule of the giant M1 transitions /4/, as will be discussed later on in that context (Sect. 2.3).

To show this, we may limit ourselves to values of q sufficiently small so that the energy integration is confined to the giant resonance region (remember that in electron scattering, $\omega \lesssim q$). For these small values of q, the form factor is also dominated by the isovector part of the electric dipole operator of (8), cf. the remark after (4), so that the latter operator may be used for the present purposes. However, the effect of retardation neglected in (8) may still be approximately taken into account by multiplying the matrix element of (8) by the elastic (L = 0) form factor $F_C(q)$ normalized to 1 at q = 0 /28/.

Using the closure procedure of (16), as further detailed in Appendix A, we obtain for the transverse E1 form factor of (34)

$$|\mathscr{T}_1^{el}(q)|^2 \cong F_C^2(q) \left\{ \alpha + \beta q^2 + \gamma q^4 \right\} \tag{36a}$$

with the ground state expectation values

$$\alpha = \frac{1}{4\pi} <0| \sum_i \tau_{3i} v_{zi} \sum_j \tau_{3j} v_{zj} |0> \tag{36b}$$

$$\beta = -\frac{\mu_p - \mu_n}{8\pi m^2} \frac{1}{3} <0| \sum_i \underline{l}_i \cdot \underline{s}_i |0> + \frac{\mu_p - \mu_n}{16\pi m} \frac{1}{3} <0| \sum_{i \neq j} \tau_{3i} \tau_{3j} \underline{v}_j \cdot \underline{r}_j \times \underline{\sigma}_j |0> \tag{36c}$$

$$\gamma = \frac{(\mu_p - \mu_n)^2}{128\,\pi} \frac{1}{m^2} <0| \sum_i \tau_{3i} (\underline{r}_i \times \underline{\sigma}_i)_z \sum_j \tau_{3j} (\underline{r}_j \times \underline{\sigma}_j)_z |0>. \tag{36d}$$

The quantity α may be computed from the well-known double-commutator sum rule (see Appendix A)

$$\sum_n \omega_n |<n| \mathcal{O} |0>|^2 = \frac{1}{2} <0| \left[\mathcal{O}^+, \left[\mathcal{H}, \mathcal{O}\right]\right] |0>, \tag{37a}$$

which holds when $|<n|\mathcal{O}^+|0>|^2 = |<n|\mathcal{O}|0>|^2$ with $\omega = E_n - E_o$, where $|n>$ is an eigenstate of the Hamiltonian with energy E_n, and $|0>$ the ground state with energy E_o. Indeed, if $\mathcal{O} = \sum_i \tau_{3i} z_i$ and $\mathcal{H}$ contains only Wigner potentials, one has

$$\sum_n \omega_n |<n| \sum_i \tau_{3i} z_i |0>|^2 = \frac{A}{2m}. \tag{37b}$$

Since the E1 giant resonance largely saturates the sum rule, one may remove its energy ω_{Res} from the sum, and obtain

$$\alpha \cong \frac{A}{m} \frac{\omega_{Res}}{16\,\pi}. \tag{37c}$$

Similarly, the coefficient γ is calculated using

$$\sum_n \omega_n |<n| \sum_j \tau_{3j} (\underline{r}_j \times \underline{\sigma}_j)_z |0>|^2 = \frac{A}{m} + \frac{1}{3m} <0| \sum_j \underline{l}_i \cdot \underline{s}_i |0>. \tag{37d}$$

The sum rules /37b,d/ hold rigorously only when Wigner forces are present. CZYZ et al. /28/ studied their breaking in the presence of Majorana forces and estimated it to be fairly small.

Finally, we calculate β taking into account only the most important parts of the correlations (i = j)

$$\beta \cong -\frac{\mu_p - \mu_n}{4\,\pi} \frac{1}{3} <0| \sum_i \underline{l}_i \cdot \underline{s}_i |0>. \tag{37e}$$

Since in the jj coupling model, $<0| \sum_i \underline{l}_i \cdot \underline{s}_i |0> = 0\ (^{16}O),\ 4(^{12}C)$, or $12(^{28}Si)$, the term β is zero or negative (although somewhat overestimated by the jj coupling model /4/) and thus, (36a) explains the form factor minimum in ^{12}C and ^{28}Si. The quantitative theory reproduces the shape of the minimum reasonably well for ^{12}C, ^{16}O, and ^{28}Si /73/.

Likewise, the dip is also furnished by the Goldhaber-Teller model (see [Ref. 13, Fig. 6.69]), the continuum shell model /72/, and the Helm model (Figure 4), but none of these latter approaches have pointed out the role that the ground-state expectation value $<0|\ \sum_j \underline{l}_j \cdot \underline{s}_j\ |0>$ plays in this context.

2.3 Higher-Multipole Giant Resonances in Electron Scattering; Sum Rules

The term "Giant Resonance" has for a long time been identified with the collective nuclear giant E1 resonance, which had been proposed in 1944 by MIGDAL /74/, had subsequently been experimentally discovered by BALDWIN and KLAIBER /1/ in 1947, and which was interpreted theoretically by GOLDHABER and TELLER /47/ in 1948. The term should now be extended to the higher-multipole nuclear resonances of collective character, whose existence was predicted theoretically in the hydrodynamical model by DANOS /75/ and by RAPHAEL et al. /12/ and in the shell model by DONNELLY and WALKER /76/, and which were discovered experimentally in 1971 in electron scattering /77, 78/ and in 1972 in proton scattering /5/. Although the existence of higher-L resonances can be expected in a rather natural way, once the existence of the E1 resonances is ascertained /15/ (which follows already from a harmonic-oscillator shell model in a straightforward way), the remarkable fact is that experimentally, these states retain their collective character, saturating a substantial fraction of corresponding sum rules /62, 79/ and not exhibiting any excessive spreading or fragmentation although they are situated in the nuclear continuum. The reason why these resonances had not been observed in the old days of photonuclear physics is that photoexcitation corresponds to the long wavelength limit /13, 76/ in which high-J transitions are strongly suppressed. To observe high multipolarities requires large values of the momentum transfer q, which can be supplied in electron or hadron scattering, for example. In addition, the monopole transitions (E0 or C0) cannot be excited in photonuclear reactions at all, due to the transverse nature of real photons.

The generalized Goldhaber-Teller model discussed in Sect. 1.4 provides a straightforward classification of all collective resonances of multipolarity J (while the symbol L is reserved for the multipolarity of the nuclear matter vibration), as shown in Table 1. In particular, the isospin character of a transition with given multipolarity J is shown to be T = 1 (isovector) for the isospin and the spin-isospin mode, or T = 0 (isoscalar) for the spin-wave and for the compressional mode. Generally speaking, isoscalar transitions correspond to collective nuclear vibrations in which protons and neutrons with a given spin direction vibrate in phase, and isovector transitions correspond to their vibrations out of phase. It should be noted that the term "compressional mode" for the type of vibration where all four nuclear fluids move in phase /12/ has been very unfortunately chosen since the in-phase vibration does not refer to the (compressional) breathing mode only, but to all other nuclear vibrations as well, only excepting L = 1.

Besides the general classification thus established, namely isovector (T = 1) E0, E1, E2, etc., and isoscalar (T = 0) E0, E1, E2, etc., one may have a further subclassification of transitions which depends on the model. For example, in the harmonic-oscillator model one may classify the transitions by the values of $\Delta\hbar\omega$ involved in the promotion of particles to higher shells, with the parity constraint only permitting even integer multiples of $\Delta\hbar\omega$ for even multipoles, and odd multiples for odd multipoles (with angular momentum and long-wavelength restrictions effectively limiting the number of allowed values for $\Delta\hbar\omega$). In the generalized Goldhaber-Teller model, the further subclassification is provided by the fact whether a spin flip (S = 1) is or is not involved in the transition (since the model is based on LS coupling). For the case S = 0, the usual EL (T = 1) and EL (T = 0) transitions appear. For the case S = 1 we find in addition to the "electric spin flip" transitions EJ (J = L) the further giant collective magnetic (spin flip) transitions MJ (J = L - 1 or J = L + 1). All these levels can be expected to be split by residual interactions which break SU4 symmetry.

Table 2. Estimated energies of collective EJ Transitions /83/

Transition	$\Delta\hbar\omega$	$\omega(A^{-1/3}\text{MeV})$	
		T = 0	T = 1
E0	2	58	178
E1	1	-	80
E2	2	58	135
E3	1	25	53
	3	107	197
E4	2	62	107
	4	152	275

Estimates for the energies of some of these collective levels have been given by BOHR and MOTTELSON /62, 80/. In Table 2, we list the energies for collective EJ transitions, J $\leq$ 4, with the values of E0 from SUZUKI /81/, and of higher multipoles from HAMAMOTO /82/. Note that from this table, the excitation energies of the isoscalar E0 and E2 states coincide.

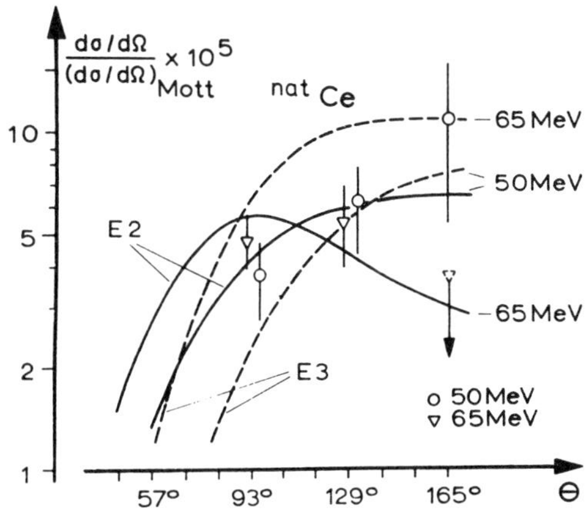

Fig. 5. Ratio of the inelastic (e,e') cross section of the resonance at 12.0 MeV in Ce to the Mott cross section, plotted for primary energies E_i of 50 and 65 MeV as a function of scattering angle. The curves show the result of DWBA calculations assuming an E2 assignment with B(E2, 0) = 2.5 · 10^3 fm^4 (solid lines) and an E3 assignment with B(E3, 0) = 4.3 · 10^5 fm^6 (dashed lines) /77/

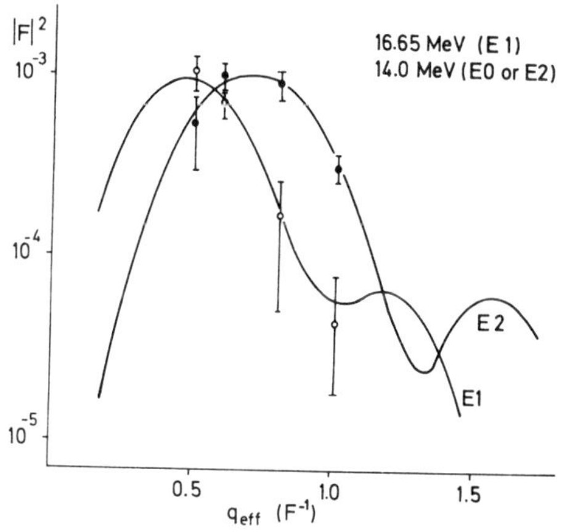

Fig. 6. The experimental form factor for the 14.0 and 16.56 MeV peaks in ^{90}Zr compared with E2 and E1 form factors /10/

In Figures 5 and 6, we present the original evidence for collective E2 transitions as discovered in inelastic electron scattering. Figure 5 shows the E2 assignment to the 12 MeV resonance in Ce /77/, and Figure 6 the E2 assignment to the 14 MeV resonance in ^{90}Zr /78/. Note that in electron scattering, E2 transitions cannot be distinguished from E0 transitions (at least in the long-wavelength limit where the E0 and the E2 cross sections start out with the same power of q; however, even for large values of q the distinction is very model-dependent and hence, not always reliable). A distinction can also be based on the fact that E0 transitions are not excited in photonuclear reactions but E2 transitions are, so that the latter levels may also be photoexcited, e.g., in (γ, n) reactions.

It may also be based on a (model-dependent) angular distribution fit in the excitation of these resonances by hadrons; but it should be noted that only isoscalar transitions get strongly excited by common hadron projectiles such as deuterons, and alpha particles /8/ (this latter fact may, however, be used advantageously in order to distinguish between isoscalar and isovector transitions, which is not the case for electron scattering, except for the spin-flip transitions). Discussions of the problem of giant monopole resonances are given in the recent review papers on the subject /10, 83-87/.

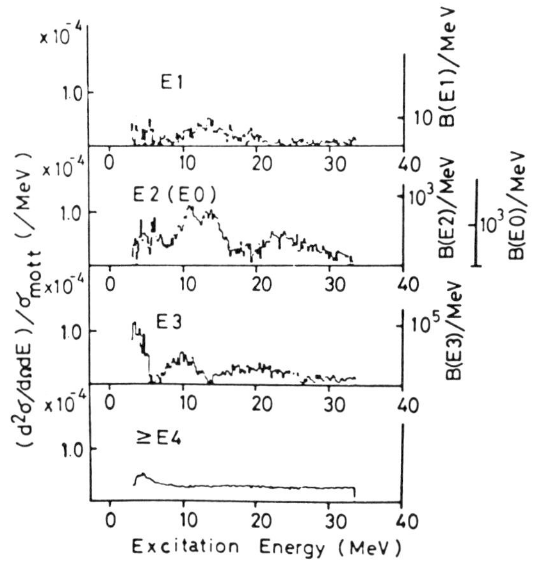

Fig. 7. The multipole components unfolded from the spectrum at 183 MeV electron energy and 35° scattering angle on ^{197}Au /87/. In the decomposition the Goldhaber-Teller model for the E1 resonance and the Tassie model for other multipoles were used /13, 21/

An unfolding method for the multipole resonances in electron scattering has been developed at the Laboratory of Nuclear Sciences at Tohoku University /85-87/, based on an unfolding method for the radiation tail devised by FRIEDRICH /88/. An example of these results, for electroexcitation of ^{197}Au, is shown in Figure 7.

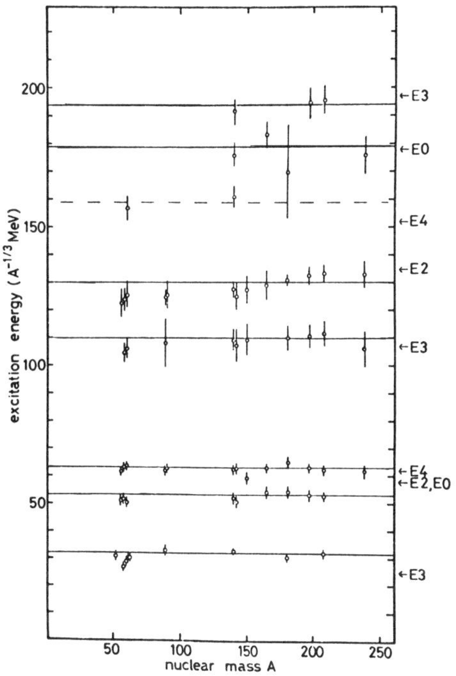

Fig. 8. Overview on existing data from (e,e') on giant resonance energies of various multipolar character /83/

Collective resonances are exhibited here separately for each transition multipolarity as a function of excitation energy. These partial cross sections have been obtained using the Goldhaber-Teller, Steinwedel-Jensen, and Tassie models, and differ somewhat depending on which model has been used.

A summary of our present experimental knowledge of higher-multipole resonances throughout the periodic table, as compiled by PITTHAN /83/, is presented in Figure 8. The horizontal lines are drawn to guide the eye, while the arrows represent the energy values from Table 2. The constancy of these excitation energies (in units $A^{-1/3}$) over a large range of A is remarkable, and confirms the general character of these nuclear excitations.

A resolution of the giant multipole peaks into the components that correspond to one given multipolarity, such as shown in Figure 7, permits a Regge pole description of the resonances in which the higher-multipole resonances are interpreted as the "Regge recurrences" of the lower-order (J = 1) resonances, hence tying higher J's to the existence of the giant dipole, and providing a unified picture of all giant resonances /15/.

In the energy (E) variable, each resonance has the form (expressed by a transition matrix element $\tilde{M}_J$, with q = momentum transfer)

$$\tilde{M}_J(q,E) = M_J(q) \frac{(\Gamma_J/2\pi)^{1/2}}{E-\omega_J+\frac{i}{2}\Gamma_J} , \tag{38}$$

Γ_J being the resonance width. Higher multipole resonance energies can be theoretically obtained by SATCHLER's formula /89/

$$\omega_J \cong C_J\ A^{-1/3} \left[L(L+3)\right]^{1/2}. \tag{39}$$

Using McVOY's transformation /90/

$$\omega_J(L) \cong \omega_J(L_E) + (L-L_E)\ \omega_J'(L_E) \tag{40}$$

with $\omega_J(L_E) = E$, (38) becomes proportional to

$$\tilde{M}_J(q,L) = M_J(q) \frac{(\hat{\Gamma}_J/2\pi)^{1/2}}{L-L_E-\frac{i}{2}\hat{\Gamma}_J} \tag{41}$$

which represents a Regge pole in the complex L plane, located at

$$L_P(E) = L_E + \frac{i}{2}\ \hat{\Gamma}_J(E), \tag{42}$$

where $\hat{\Gamma}_J(E) = \Gamma_J/\omega_J'(E)$. Equation (42) plotted as a function of E describes a Regge trajectory in the L plane. These trajectories are shown for the case of ^{16}O in

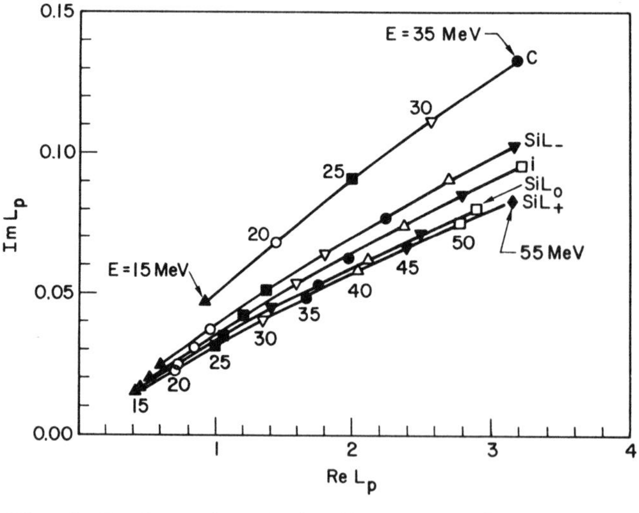

Fig. 9. Trajectories of the five Regge poles generating the giant multipole resonances of the generalized Goldhaber-Teller model ($L_o \to L = J$; $L_\pm \to L = J \pm 1$) /15/

Figure 9 /15/. If the transition amplitude $\mathcal{A}_L(E)$ is expressed by (41), the important factors are

$$\mathcal{A}_L(E) = F(L)\,\frac{P_L(\cos\Theta)}{L-L_p(E)}. \qquad (43)$$

A Watson transformation evaluation of (43) gives for the total amplitude of the resonances

$$\mathcal{A}(E) = \sum_L \mathcal{A}_L(E) = F(L_p)\,\frac{\pi P_{L_p}(-\cos\Theta)}{\cos\pi(L_p+\frac{1}{2})}\,. \qquad (44a)$$

The asymptotic form

$$P_L(\cos\theta) \sim \sqrt{\frac{1}{2\pi L\sin\theta}}\left\{\exp i\left[(\nu+\tfrac{1}{2}) - \tfrac{\pi}{4}\right] + \exp - i\left[(\nu+\tfrac{1}{2})\theta - \tfrac{\pi}{4}\right]\right\} \qquad (44b)$$

suggestively /15/ represents two attenuated surface waves propagating over the nucleus in opposite senses, plus an infinite number of additional waves that already have encircled the nucleus m times previously. One would argue a) a given giant resonance mode is described by the motion of a Regge pole in the L plane which generates in successsion the dipole as it passes L = 1, the quadrupole as it passes L = 2, etc., and b) the condition for an L resonance to occur is that the wavelength of one of the surface waves be such that $L + \frac{1}{2}$ wavelengths fit over the nuclear circumference.

At that energy, constructive interference occurs since the wave loses a quarter-wavelength as it passes the caustic points at the north and the south pole of the (spherical) nucleus, and the repeated revolutions of the surface wave result in a resonant reinforcement of the amplitude.

From (44a), using the asymptotic form of $P_L(-\cos\Theta)$ (44b) the wavelenght of the surface waves is found as

$$\lambda(E) = 2\pi R/(\mathrm{Re}L_p + 1/2), \qquad (45)$$

where R is the nuclear radius. Their attenuation is given by exp $(-\Theta/\overline{\Theta})$ in amplitude, with a decay angle

$$\overline{\Theta}(E) = 1/\mathrm{Im}L_p(E). \qquad (46)$$

Finally, the phase velocity of the collective surface waves,

$$C(E) = RE/(\mathrm{Re}L_p(E) + 1/2) \qquad (47)$$

as a function of projectile energy E can be represented in the form of dispersion curves as shown in Figure 10. Here, we denote be L_o: L = J; by $L_\pm$: L = J ± 1.

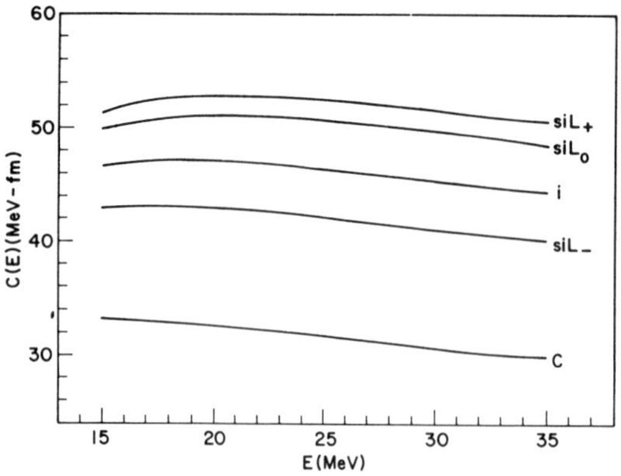

Fig. 10. Phase velocities of surface waves vs energy transfer /15/

Various types of sum rules have played an important role in the analysis of inelastic electron scattering by nuclei. Sum rules were initially formulated for the scattered electron spectrum summed over the energy ω transferred to the nucleus, at a specific electron scattering angle Θ and a specific initial energy E_i, i.e., for the angular distribution of the scattered electrons /91/. A theoretical summation can be made in this case only in a rather crude approximation. It was later proposed to carry out the summation over ω at fixed values E_i and q, where $q = |\vec{q}|$ is the momentum transferred by the electron to the nucleus, or, in another variant, at fixed values of q and Θ /91, 92/. In these cases, particularly in the latter, the theoretical summation can be carried out more accurately.

Details of the technical ingredients entering the sum rules are given in Appendix A. Here we briefly mention the fact that the sum rule techniques related to the generalized Goldhaber-Teller model (in the sense that they are saturated by the states which are featured in this model) suggest a possible method to tackle all the reactions in which giant resonance phenomena are involved (nucleon-nucleus scattering, heavy ion scattering /93/, pion-nucleus scattering, etc.). These techniques provide a parameterization in terms of centroids, widths, etc., of the strength of the resonant operators of various multipolar characters. In particular, besides the electric sum rules, magnetic dipole (M1) transitions /94/ which have been studied in numerous nuclei using inelastic electron scattering, have been analyzed using KURATH's sum rule /95/. With the modern electron accelerators, a series of high-multipolarity magnetic transitions (as high as M6, M8, M12, M14) have recently been identified /96-98/. In Figure 11, the interpretation of various strong transitions in ^{58}Ni as M8 is shown by shell-model fits to the measured transverse form factors /97/. Such transitions can be analyzed in terms of an ML (T = 1) sum rule derived in the long-wavelength limit /99/. This derivation starts from the double-commutator identity (in absence of exchange, spin-orbit, and velocity-dependent potentials)

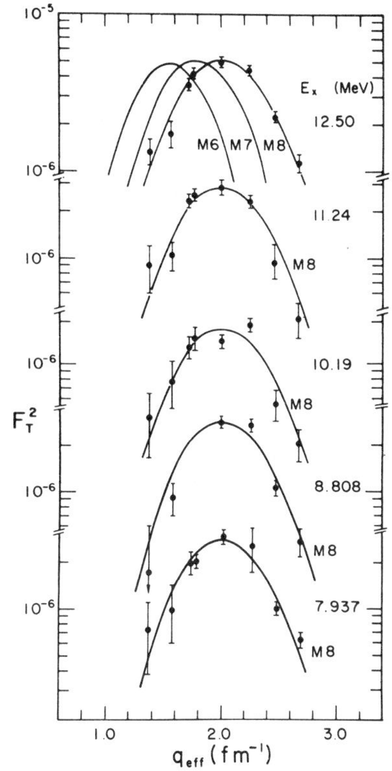

Fig. 11. Measured /97/ transverse form factor F_T^2 vs effective /13/ momentum transfer q_{eff} for several transitions in ^{58}Ni

$$\left[\hat{Q}^{+}_{1LM}, \left[H, \hat{Q}_{1LM}\right]\right] = \left(\frac{1}{4\pi}\right) \sum_{i=1}^{A} |\underline{\nabla}_i \, q^{i}_{1M}|^2 \tag{48}$$

with the transitions operator written in the form

$$\hat{Q}_{1LM} = \frac{1}{2} \sum_{i=1}^{A} q^{i}_{1M} \, \tau_{3i} \, . \tag{49}$$

For the ML transitions considered here, we may just retain the dominant spin-flip part

$$q^{i}_{1M} = i^L \, (\gamma/2m) \, \underline{\sigma}_i \cdot \underline{\nabla}_i \times j_L \, (qr_i) \, \underline{Y}^{M}_{LL} \, (\hat{r}_i), \tag{50}$$

where $\gamma = \mu_p - \mu_n$, since orbital magnetic current contributions are known to be small. From the expectation value of the double commutator in the state $|J_oM_o\rangle$, one obtains the energy-weighted sum rule for the transverse form factors $F_T^2(q)$

$$\sum_n \omega_n F_T^2(q) = \frac{1}{8m} \frac{\hat{L}^2}{\hat{J}_o^2} \sum_M \langle J_oM_o| \sum_{i=1}^{A} |\underline{\nabla}_i \, q^{i}_{M}|^2 \, |J_oM_o\rangle, \tag{51}$$

where n stands for all quantum numbers in the final states except for J_F and M_F (the sum over the latter having been carried out). This equation represents a sum rule which is valid for all values of the momentum transfer q. If the right-hand side is evaluated in the long-wavelength limit, we obtain

$$\sum_n \omega_n F_T^2(q) = \frac{1}{4\pi} \frac{A}{8m} \left(\frac{\gamma}{2m}\right)^2 \frac{q^{2L}}{[(2L+3)!!]^2} \frac{(L-1)(L+1)}{2L-1} \langle r^{2L-4} \rangle_0 \tag{52}$$

in terms of the ground-state average of r^{2L-4}. Here, L is the multipolarity of the magnetic (ML) transition $|J_0 M_0\rangle \to |JM\rangle$. This sum rule was derived by MONTGOMERY and ÜBERALL /99/ and independently by TRAINI /99/, who also obtained spin-orbit coupling and residual-force corrections. For the high-L transitions, it allows us to put stringent limits on the shape of the nuclear surface, due to the extreme sensitivity of high moments $\langle r^{2L-4} \rangle_0$ to the nuclear surface.

Sum rules in general provide a method by which the degree of exhaustion of the transition strength in a sum over the known nuclear levels of a given multipolarity may be determined, and hence they furnish us with an indication whether a search for more of these levels might be fruitful. Higher-multipole sum rules have previously been known for EL transitions /79/; they represent generalizations of the well-known Thomas-Reiche-Kuhn E1 sum rule. For example, there has recently been a widespread interest in the location and in the saturation of the E2 giant resonance /100/. The latter has been experimentally studied, e.g., in the Ni isotopes /101/ using a new method in which the yields of (γ,p) and (γ,α) reactions are compared when initiated either by real photons, or by the virtual photons of the (e, e') reaction. The method is based on the larger predominance of E2 over E1 photons in the virtual photon spectrum. It is illustrated in Figure 12 /79/.

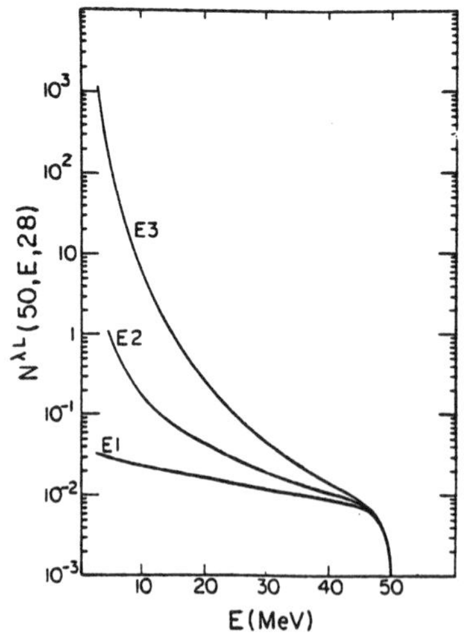

Fig. 12. The E1, E2, E3 DWBA virtual photon spectra for a 50 MeV electron scattered by a Nickel nucleus /79/

The EL (L ≧ 2) energy-weighted sum rules used to analyze the data have been given in BOHR and MOTTELSON /62/. They are as follows, when expressed in terms of photonuclear absorption cross sections /79/:

a) *isoscalar E2*

$$\int \frac{\sigma d\omega}{\omega^2} = \frac{\pi^2 e^2}{3m}\frac{Z^2}{A}\langle r^2\rangle_0 \tag{53a}$$

b) *isovector E2*

$$\int \frac{\sigma d\omega}{\omega^2} = \frac{\pi^2 e^2}{3m}\frac{NZ}{A}\langle r^2\rangle_0 \tag{53b}$$

c) *isoscalar E3*

$$\int \frac{\sigma d\omega}{\omega^4} = \frac{4\pi^2 e^2}{225m}\frac{Z^2}{A}\langle r^4\rangle_0 \tag{53c}$$

d) *isovector E3*

$$\int \frac{\sigma d\omega}{\omega^4} = \frac{4\pi^2}{225m}\frac{NZ}{A}\langle r^4\rangle_0 . \tag{53d}$$

Contributions to the E2 sum rule, coming from the spin-flip terms of the E2 operator and not included in (53), have also been obtained theoretically /94/ in order to describe several of the electric spin-flip transitions which are known by now in some light nuclei /102/. Adopting a shell-model Hamiltonian with spin-orbit coupling (but neglecting residual forces between nucleons in unfilled shells)

$$H - \sum_{i=1}^{A} \frac{p_i^2}{2m} + V(r_1) + a\, \underline{l}_i \cdot \underline{s}_i$$

one obtains the <u>isovector</u> sum rule valid in the long-wavelength limit

$$\sum_n \omega_n \left| \mathcal{T}_2^{el} \right|^2 \cdot \frac{2}{3}\left(\frac{15}{\hat{j}_o}\right)^2 = q^2(S_j + S_2 + S_a). \tag{54}$$

It contains the contributions from the current transitions /94/,

$$S_j = \frac{25}{16\pi}\left(\frac{\omega}{q}\right)^2 \frac{1}{m}\langle r^2\rangle_0 , \tag{55a}$$

which represent the conventional [Cf. (53b)] E2 sum rule; contributions from the 2 ħω spin-flip transitions,

$$S_2 = \frac{25}{6}\frac{\gamma^2}{4}\left(\frac{q}{2m}\right)^2 \frac{1}{m}\left\{\langle r^2\rangle_0 + \frac{1}{10}\langle r^2 \underline{\sigma}\cdot\underline{l}\rangle_0\right\} \tag{55b}$$

and finally contributions due to the effect of spin-orbit splitting

$$S_a = -\frac{5a}{9\pi}\frac{\gamma^2}{4}\frac{q}{2m}\left\{\frac{3}{4}<r^4>_0 + <r^4\underline{\sigma}\cdot\underline{l}>_0 + \right.$$

$$\left. + \left(\frac{3\pi}{2}\right)^{1/2} <r^4 \left[Y_2 \times [\sigma \times l]_2\right]_0 >_0 \right\}. \tag{55c}$$

As far as the isospin structure of the excitation operators is concerned, we note that due to the fact that one is not necessarily in the long-wavelength limit in electron scattering, one can here have isoscalar electric dipole resonances also. For real photons, the E1 operator is purely isovector (because of center-of-mass considerations). Thus, a decay (T = 0)E1 → (T = 0) by real photon emission is considered isospin forbidden and occurs only through the admixing of T = 1 components. However, the isoscalar operator for virtual dipole photon absorption, such as it occurs in inelastic (longitudinal) electron scattering, is modified by retardation effects (Bessel functions) so that it is not annihilated by center-of-mass effects.

Collective-model estimates of transition strengths, as well as sum rules for isoscalar transition densities have been obtained, e.g., for monopole isoscalar vibrations and for isoscalar dipole resonances /103/, which have been measured in recent electron scattering experiments /104/.

3. Giant Resonances in Muon Capture

Giant resonance excitation in muon capture has been considered in previous review papers /23, 34, 105/, so that in the present context, we shall after a brief introduction limit ourselves to the discussion of more recent progress. The theoretica expressions for the muon capture rate [Ref. 23, Eq. (1), Ref. 34, Eq. (9a)] mainly depend on the three (squared and averaged) matrix elements M_V^2, M_A^2 and M_P^2 defined by

$$M_V^2 = \frac{1}{\hat{J}_o^2} \sum_{M_o} \sum_{n,M_n} \left(\frac{\nu_{on}}{m_\mu}\right)^2 \int \frac{d\hat{\nu}}{4\pi} |\langle n| \sum_i \tau_i^- \exp -i\underline{\nu}_{on}\cdot\underline{r}_i |0\rangle|^2 \tag{56a}$$

$$M_A^2 = \frac{1}{\hat{J}_o^2} \sum_{M_o} \sum_{n,M_n} \left(\frac{\nu_{on}}{m_\mu}\right)^2 \int \frac{d\hat{\nu}}{4\pi} |\langle n| \sum_i \tau_i^- \frac{\underline{\sigma}_i}{\sqrt{3}} \exp -i\underline{\nu}_{on}\cdot\underline{r}_i |0\rangle|^2 \tag{56b}$$

$$M_P^2 = \frac{1}{\hat{J}_o^2} \sum_{M_o} \sum_{n,M_n} \left(\frac{\nu_{on}}{m_\mu}\right)^2 \int \frac{d\hat{\nu}}{4\pi} |\langle n| \sum_i \tau_i^- \underline{\sigma}_i\cdot\hat{\nu} \exp -i\underline{\nu}_{on}\cdot\underline{r}_i |0\rangle|^2, \tag{56c}$$

where

$|0\rangle$ = initial nuclear state with spin (J_oM_o),

$|n\rangle$ = final nuclear state,

$\underline{\nu}_{on}$ = neutrino momentum corresponding to $|0\rangle \to |n\rangle$,

m_μ = muon rest mass.

Other matrix elements corresponding to relativistic corrections /22/ are relevant mainly in specific partial transitions.

3.1 Giant Dipole Excitations

If the Hamiltonian of the nuclear target is SU4 invariant, then by taking matrix elements of the commutator

$$\frac{1}{2}\left[\sum_i \tau_i^- \exp -i\underline{\nu}\cdot\underline{r}_i \, , \sum_j \underline{\sigma}_j \tau_{3j}\right] = \sum_i \underline{\sigma}_i \tau_i^- \exp -i\underline{\nu}\cdot\underline{r}_i \tag{57}$$

between a scalar supermultiplet $|0\rangle$ and a state $|n\rangle$ belonging to a vector supermultiplet one may deduce the relation /31/

$$M_V^2 = M_A^2 = M_P^2 \tag{58}$$

between the three principal matrix elements entering in the capture rate. Of these, M_V^2 may be known quite accurately from experiments because in the unretarded dipole (UD) approximation, see (8), it may be related typically to within 10% accuracy to the photoabsorption cross section $d\sigma(\omega)/d\omega$ through the commutator (14). This relation, found by FOLDY and WALECKA /31/, is

$$\left(M_V^2\right)_{UD} = \frac{m_\mu^2}{2\pi^2\alpha}\left(\frac{\omega_m}{m_\mu}\right)^4 \int_0^{\omega_m} \frac{1}{\omega}\left(\frac{\omega_m - \omega}{\omega_m}\right)^4 \frac{d\sigma(\omega)}{d\omega}\, d\omega, \tag{59}$$

where the cutoff energy is $\omega_m = m_\mu - \varepsilon_\mu$ with ε_μ = (binding energy of the muon in the S orbit) + (neutron-proton mass difference) - (Coulomb energy difference between isobaric states). In the SU4 limit, the knowledge of M_V^2 would thus be sufficient for the calculation of total muon capture rates; but if SU4 breaking effects in the excited states are also considered (/106/, see Appendix B), using an energy-weighted sum rule, the splitting between the spin-flip and non-spin-flip components of the resonances enters as a new parameter, as discussed, e.g., in /34/, which may be taken into account in a phenomenological way.

While the relation of (58), when derived using the commutator expressions of (57), implies the use of the impulse approximation, one may derive the relation $M_V^2 = M_A^2$ in a more general way on the basis of current algebra. We still assume that the SU4 group, generated by the moments of the physical currents, is a symmetry group of the nuclear Hamiltonian, so that the excited states may be grouped into supermultiplets and the ground state is a scalar supermultiplet.

It has been emphasized by RADICATI /35/ that the relevant commutators may be written as equal-time commutators of the nuclear currents $J^{nuc}_{weak,\mu}(\underline{r})$ (which are not necessarily 1-body currents as in (11) but physical currents /107/):

$$\left[\int d^3r\, J^{\pm}_{ax,k}(\underline{r}), \int d^3r' \exp -i\underline{\nu}\cdot\underline{r}'\, \rho^3_{vec}(r')\right]_{t=t'=0} = \int d^3r \exp -i\underline{\nu}\cdot\underline{r}\, J^{\pm}_{ax,k}(\underline{r}) \tag{60a}$$

$$\left[\int d^3r\, \rho^i_{vec}(\underline{r}), \int J^j_{ax,\mu}(\underline{r}')\, d^3r'\right]_{t=t'=0} = i\varepsilon_{ijk} \int J^k_{ax,\mu}(\underline{r})\, d^3r \qquad (i,j,k = 1,2,3) \tag{60b}$$

$$\left[\int J^i_{ax,1}(\underline{r})\, d^3r, \int J^j_{ax,1}(\underline{r})\, d^3r'\right]_{t=t'=0} = i\varepsilon_{ijk} \int \rho^k_{vec}(\underline{r})\, d^3r. \tag{60c}$$

The physical currents may include exchange effects if one assumes that they obey the same algebra as the free quark model /108/. Further details will be given in Appendix B, with the result $M_V^2 = M_A^2$ following from (60). It is clear that such a procedure is quite general in view of the fact that in the photabsorption sum rule, (13), exchange potentials give contributions of 40-100% /44, 109/ while our use of (60) automatically includes exchange effects.

3.2 Muon Capture in N>Z Nuclei

A relation between dipole excitations in photoabsorption and in muon capture may also be established for N>Z nuclei in analogy with the procedure of FOLDY and WALECKA /31/. This can be proved /110/ as follows: With a ground state

$$|0\rangle = |T_0, T_3 = -T_0\rangle \tag{61a}$$

and the excited dipole states in the same nucleus,

$$|n'\rangle = |T_0 + 1, T_3 = -T_0\rangle \tag{61b}$$

and in the neighboring nucleus

$$|n\rangle = |T_0 + 1, T_3 = -T_0 - 1\rangle \tag{61c}$$

and noting that

$$T^-|0\rangle = 0 \tag{61d}$$

one immediately obtains using (14)

$$\frac{1}{\hat{J}_o^2} \sum_{M_o} \sum_{n,M_n} |\langle n| \sum_i \tau_i^- \underline{r}_i |0\rangle|^2 = $$
$$= \frac{1}{2} (T_o + 1) \frac{1}{\hat{J}_o^2} \sum_{M_o} \sum_{n,M_n} |\langle n'| \sum_i \tau_{3i} \underline{r}_i |0\rangle|^2 . \tag{62}$$

This permits us to write $(M_V^2)_{UD}$ as an integral over $d\sigma^{T_o+1}(\omega)/d\omega$, which is the photoabsorption cross section $(T_o \rightarrow T_o + 1)$ at excitation energy ω

$$\left(M_V^2\right)_{UD} = \frac{m_\mu^2}{2\pi^2\alpha} \left(\frac{\omega_m}{m_\mu}\right)^4 (T_o + 1) \int_0^{\omega_m} \frac{1}{\omega} \left(\frac{\omega_m - \omega}{\omega_m}\right)^4 \left(d\sigma^{T_o+1}(\omega)/d\omega\right) d\omega . \tag{63}$$

To evaluate this expression, we need an estimate of the (energy-weighted) $T_o + 1$ photoexcitation cross section. This may be obtained from the experimental total energy-weighted photoexcitation cross section using relations derived by FALLIEROS and GOULARD /111/ and HAYWARD et al. /112/. This procedure has been applied to muon capture in several light nuclei, in order to assess the importance of the

Table 3. Experimental total muon capture rates in several N≧Z nuclei, compared with calculated dipole capture rates using (63): ^{6}Li from /110/, other nuclei from /113/

Rate	Nucleus						
$(10^5\ s^{-1})$	^{6}Li	^{9}Be	^{14}N	^{19}F	^{24}Mg	^{27}Al	^{28}Si
Λ_{tot}	0.061	0.10	0.64	1.4	4.8	7.0	8.5
Λ_{dip}	0.013	0.034	0.46	1.3	2.4	4.6	5.7

giant dipole contribution to the total capture rate /113/, with results given in Table 3. The table shows that, apparently with the exception of ^{6}Li and ^{9}Be (we note that here hyperfine effects might in principle be important), the dipole contribution is quite predominant in light nuclei for N>Z as well as N = Z nuclei; this is known not to be the case in heavier nuclei /34/.

3.3 Improvement of the Closure Approximation

The conventional closure approximation /114/ consists in approximating an energy-weighted sum rule in the expression for total muon capture rates, see (56), such as e.g., given in (15), by a non-energy-weighted one, i.e., (16) with $q = 0$, via approximating the neutrino momentum ν by a mean value $\bar{\nu}$, and thus taking the energy weight out of the sum. The resulting expression contains a two-body operator, to be evaluated in the ground state, and thus depends on the ground-state nucleon-nucleon correlations. However, this procedure suffers from uncertainties due to the degree of arbitrariness involved in the value of $\bar{\nu}$, so that it appears reasonable to Taylor-expand the energy dependence in the partial rate and then summing to obtain zero and first-order energy-weighted sum rules, cf. [Ref. 34, Eq. (21)]. In this way, GOULARD and PRIMAKOFF /115/ found the expression for the capture rate

$$\Lambda = KZ^4_{eff}\left(1 - \frac{\varepsilon_\mu}{m_\mu}\right)^2 \left[1 - 2\,\frac{m_\mu - \varepsilon_\mu}{2m}\right] \cdot \left\{1 + \frac{A}{2Z}\,\beta_o - \frac{A-2Z}{2Z}\,\beta_1 - \left(\frac{A-Z}{2A} + \frac{|A-2Z|}{8ZA}\right)\beta_2\right\} \tag{64a}$$

with

$$K = \left(G_V^2 + 3G_A^2 + G_P^2 - 2G_PG_A\right)\frac{\alpha^3 m_\mu}{2\pi^2}\ , \tag{64b}$$

where G_V, G_A, and G_P are the effective muon capture coupling constants obtained, e.g., in /34/. Here, β_o, β_1, and β_2 are treated as independent of A and Z, so that (64a) represents a three-parameter fit to the experimental data on the total muon capture rates (note its explicit independence of the average neutrino momentum). With the values

$$\begin{aligned} \beta_o &= -0.030, \\ \beta_1 &= -0.25, \\ \beta_2 &= \ \ 3.24, \end{aligned} \tag{64c}$$

one obtains the capture rates shown in Table 4, together with the measured rates. This comparison shows the reliability of the parameterization given by (64a). The fit is characterized by the mean absolute deviation $|(\Lambda_{exp} - \Lambda_{fit})/\Lambda_{exp}| = 5.6\%$ averaged over 57 elements $8 \leq Z \leq 92$. The good quality of the fit indicates that β_o, β_1, and β_2 are constant throughout the periodic table.

ROSENFELDER /116/ has improved the closure approximation further by introducing two (or more) mean excitation energies. A systematic expansion scheme based on the

Table 4. Total muon capture rates in medium heavy and heavy nuclei as fit by (64a), and as measured experimentally /115/

Element	Λ_{fit} (10^6 s^{-1})	Λ_{exp} (10^6 s^{-1})
$_8O$	0.1151	0.0974 ± 0.0031
$_{16}S$	1.244	1.338 ± 0.007
$_{20}Ca$	2.46	2.45 ± 0.02
$_{24}Cr$	3.06	3.29 ± 0.04
$_{30}Zn$	5.48	5.74 ± 0.04
$_{42}Mo$	9.55	9.22 ± 0.06
$_{48}Cd$	10.66	10.62 ± 0.10
$_{54}Ba$	10.40	10.18 ± 0.10
$_{64}Gd$	11.89	12.09 ± 0.16
$_{73}Ta$	12.74	12.86 ± 0.13
$_{82}Pb$	13.05	13.02 ± 0.11
$_{92}U$	11.5	11.0 ± 0.5

energy moments is developed, by replacing the previously mentioned Taylor expansion of (15) by a superposition of expressions (15) with f(n) evaluated at several different mean energies $\bar{\omega}_n$; the latter can be calculated in terms of ground-state moments of the Hamiltonian. When the 0th, 1st, 2nd, and 3rd moments are calculated for ^{16}O between Hartree-Fock ground-state wave functions using Skyrme-type two-body forces /117/, a considerable reduction of total rates is obtained compared to the rates of the independent-particle model /34/, but the results are still 10-60% higher than the experimental value. Discrepancies in photoabsorption sum rules /109, 118/ also indicate the need for ground-state correlations.

3.4 Shell-Model Calculations

As a result of the previously mentioned work by ROSENFELDER /116/ we know that it is unlikely that an effective interaction based on Hartree-Fock wave functions can correctly give the total muon capture rate and at the same time furnish a reasonable single particle spectrum. This result is quite important in view of the fact that in general, the parameters of the effective Skyrme interaction are determined by requiring that they accurately reproduce the total binding energies and charge radii of magic nuclei in spherical self-consistent calculations. It is shown that in general many parameter sets can satisfy these requirements /119/; they differ essentially by the single particle spectra they give (they may, e.g., describe well the single particle energies of deeply bound states but fail to reproduce the spectrum around the Fermi surface, or vice versa). Thus, Rosenfelder's result means that a Skyrme parameterization of the effective interaction (based on Hartree-Fock theory) is not completely adequate, indicating a need to introduce dynamical correlations in the initial and final states in order to predict total muon capture rates as low as they are given by the experiments. Such a need for correlations had already been found to exist within the framework of the closure approximation by WALKER and McCARTHY /120/. These authors, when including correlation effects, obtained results in qualitative agreement with experiments. A reduction of calculated capture rates was also obtained in the nuclear shape coexistence model for ^{16}O /34/ (which explicitly contains ad hoc ground-state particle-hole (p-h) correlations); however, the role of excitations of positive-parity states (e.g., allowed transitions, quadrupole transitions, etc.) was not considered in detail in that context.

As was discussed in /34/, there thus exists a clear need to consistently introduce within the shell-model picture nuclear correlations in the ground state and in the excited states (of course, in a sum rule approach or in closure only correlations in the ground-state are relevant). An important step in this direction has recently been taken by ERAMZHYAN et al. /121/ including subspaces of $2\hbar\omega$ (and an important part of $3\hbar\omega$) shell-model excitations for the positive (negative) parity nuclear levels in ^{16}O. The ground-state wave function is constructed within $0\hbar\omega$ and $2\hbar\omega$ excitations. For negative parity states, the 3p-3h components in the $3\hbar\omega$ model space are disregarded and only certain 2p-2h ($3\hbar\omega$) configurations, which are most strongly coupled to the ground state, are included. The NN interaction is taken as given by the TABAKIN /122/ nonlocal separable interaction. Contrary to other model calculations, correlations in the ground state are here taken into account consistently with those in the excited states. ERAMZHYAN's work therefore indicates in a reliable way the role of explicit 2p-2h components in the final negative-parity nuclear states. Their effect is different for individual groups of transitions, as follows: in the excitation of the ^{16}N bound states, they decrease the transition rates. As to the excitations of the resonant dipole states, they may provide an enhancement of the rates. As far as excitation of positive parity states is concerned, quadrupole excitations are not found to be important (their contribution is only 10% of the total capture rate). The results of this calculation within the subspace considered are, for the summed rate of negative parity transitions $103.28 \cdot 10^3 s^{-1}$, for the positive parity transitions $9.28 \cdot 10^3 s^{-1}$, i.e., a combined rate of $112.56 \cdot 10^3 s^{-1}$. This compares favorably with an experimental total rate of $\sim 100 \cdot 10^3 s^{-1}$ /34/; the result of the naive shell model is $157 \cdot 10^3 s^{-1}$ /34/.

4. Resonance Excitation by Neutrinos

Neutrino-induced reactions are capable in principle of providing us with information concerning weak interactions (e.g., the V-A structure (11), nucleon weak form factors, existence of the vector boson) which cannot be obtained otherwise. The feasibility of high-energy neutrino experiments was pointed out by PONTECORVO in 1960 /123/ employing neutrinos emitted by pions. Many well-known experiments with neutrinos have been carried out since then, mainly concentrating on the elementary particle aspects of neutrino interactions. They have lately included also the study of neutral-current-type neutrino interactions, as predicted by the unified theories of weak and electromagnetic interactions proposed by WEINBERG /124/ and by SALAM and WARD /125/ known as the Weinberg-Salam model.

Here, we are interested in neutrino-induced reactions as probes of nuclear structure, and in particular of giant resonance states, suggested as early as 1962 by BELYAEV /126/. Although of small cross section, neutrino reactions represent weak interactions for which the momentum transfer q varies (unlike in muon capture), so that these processes are somewhat similar to electron scattering as far as kinematics is concerned (while muon capture, with a fixed 4-momentum transfer, may rather be related to photoabsorption processes as we have seen in the preceding section). For example, with suitable kinematics (i.e., forward neutrino scattering and relativistic outgoing leptons) one may obtain $q_\lambda^2 = 0$.

4.1 Charged and Neutral Current Interactions

For neutrino interactions with initial energies beyond several hundreds of MeV, the cross sections may contain contributions from final-state nucleons with relativistic energies, so that nonrelativistic nuclear currents as given in (11b-e) may be unsuitable especially at high momentum transfers. In a relativistic framework, the usual charge-changing semileptonic weak Hamiltonian is given by

$$H_W = \frac{i}{\sqrt{2}}\left[\bar{\Psi}_e\gamma_\lambda(1+\gamma_5)\Psi_{\nu_e} + \bar{\Psi}_\mu\gamma_\lambda(1+\gamma_5)\Psi_{\nu_\mu}\right] J^{nuc^\pm}_{weak,\lambda} , \qquad (65)$$

where the weak nuclear current has the form of (11a), with single-nucleon matrix elements

$$<\underline{p}'|\, J^{nuc^{\pm}}_{vec,\lambda}\, |\underline{p}> = i\bar{u}(\underline{p}')\left[F_1\,\gamma_\lambda + F_2\,\sigma_{\lambda\sigma}\,q_\sigma\right]\tau^{\pm}u(\underline{p}) \tag{66a}$$

$$<\underline{p}'|\, J^{nuc^{\pm}}_{ax,\lambda}\, |\underline{p}> = i\bar{u}(\underline{p}')\left[F_A\,\gamma_5\,\gamma_\lambda - iF_{PS}\gamma_5 q_\lambda\right]\tau^{\pm}u(\underline{p}), \tag{66b}$$

where the 4-momentum transfer $q_\lambda = p_\lambda - p'_\lambda$. As $q_\lambda^2 \to 0$, these expressions reduce in the nonrelativistic limit to the previous expressions of (11), with an additional pseudoscalar term, with the $q_\lambda^2 = 0$ limits

$$F_1(0) = g_V \qquad F_A(0) = g_A$$

$$F_1(0) + 2mF_2(0) = g_V(\mu_p - \mu_n) \tag{67}$$

$$F_{PS}(q_\lambda^2) = \frac{2m\,F_A(q_\lambda^2)}{q_\lambda^2 + m_\pi^2}.$$

As far as neutrino interactions are concerned, the Weinberg-Salam model contains an additional neutral current interaction

$$H_{WS} = \frac{i}{\sqrt{2}}\left[\bar{\Psi}_{\nu_e}\,\gamma_\lambda(1+\gamma_5)\,\Psi_{\nu_e} + \bar{\Psi}_{\nu_\mu}\,\gamma_\lambda(1+\gamma_5)\,\Psi_{\nu_\mu}\right] J_\lambda^{(0)}, \tag{68}$$

where the neutral hadronic current is

$$J_\lambda^{(0)} = J^{nuc^3}_{weak,\lambda} - 2\,\frac{g_V}{e}\,\sin^2\Theta_W\, J^{nuc}_{em,\lambda}\,. \tag{69}$$

Here $J^{nuc^3}_{weak,\lambda}$ is obtained from $J^{nuc^{\pm}}_{weak,\lambda}$ by isospin rotation. The angle Θ_W is known as the Weinberg angle, with a currently accepted value /127/ $\sin^2\Theta_W = 0.23$. Nuclear excitations by neutrinos, using the elementary interaction Hamiltonians of (65) or (68) together with the single-nucleon matrix elements of the currents such as (66), may then be calculated by expressing the nucleon 4-spinors $u(\underline{p})$ by 2-spinors corresponding to "large" and "small" components in the well-known way /13/

$$u(\underline{p}) = \left(\frac{E_p + m}{2m}\right)^{1/2} \begin{pmatrix} I \\ \underline{\sigma}\cdot\underline{p}/(E_p + m) \end{pmatrix} \chi. \tag{70}$$

The operator acting on the 2-spinor χ is then allowed to act on the individual nucleon degrees of freedom in the nuclear wave function, in the spirit of the impulse approximation. For example, one obtains thus for the change-exchange Hamiltonian the nonrelativistic expression

$$H = (\chi_p^+\,\Psi_1^+\,\mathcal{O}\,\Psi_\nu\,\chi_n), \tag{71a}$$

where

$$\mathcal{O} = \left[F_1 + F_A \underline{\sigma} \cdot \underline{\sigma}^N + F_{PS} (\underline{q}/2m) \cdot \underline{\sigma}^N \gamma_4 \right.$$

$$\left. + \, i \, (F_1 + 2m \, F_2) \, (\underline{q}/2m) \cdot (\underline{\sigma} \times \underline{\sigma}^N)\right] (1 + \gamma_5)/\sqrt{2}. \qquad (71b)$$

χ are nucleon Pauli spinors and $\underline{\sigma}^N$ Pauli matrices; Ψ_ν is the initial neutrino spinor and Ψ_1 the final charged lepton spinor.

4.2 Neutrino Cross Sections

The kinematics of the neutrino-nuclear reaction

$$\nu_1 + A_Z \rightarrow A_{Z+1}^{(*)} + 1^-, \qquad (72)$$

where 1 = electron or muon, gives for the nuclear recoil $\underline{q} = \underline{\nu} - \underline{1}$ ($\underline{\nu}$, $\underline{1}$ being three-momenta of ν and 1), and for the lepton energy $E_1 = \nu - \omega$ where ω is the final nuclear energy measured from the ground-state of the initial nucleus. Conserved vector current hypothesis suggests $F_1(q^2) = g_V F_p(q^2)$, the proton charge form factor, for which one may take

$$F_p(q^2) = 1/\left[1 + (q^2/M^2)\right]^2, \qquad (73)$$

where

$$M \cong 0.90 \text{ m} \cong 840 \text{ MeV},$$

corresponding to the proton radius 0.8×10^{-13} cm. The other form factors may be assumed to be the same (except for the pseudoscalar form factor).

If terms of order q/2m are disregarded in (71b), one obtains a cross section /66/ for the reaction of (72) corresponding to the upper signs (or the antineutrino reaction corresponding to the lower signs)

$$\frac{d\sigma}{d\Omega_1} = \frac{1E_1}{4\pi^2} \frac{1}{\hat{J}_o^2} \sum_{M_o M_f} \left\{ F_1^2 (1 + \hat{\nu} \cdot \tilde{1}) |\mathcal{M}|^2 - \right.$$

$$- 2F_1F_A \, [\text{Re} \, \mathcal{M}^* \underline{\mathcal{M}} \cdot (\hat{\nu} + \tilde{1}) \pm \text{Im} \, \mathcal{M}^* \, \underline{\mathcal{M}} \cdot (\hat{\nu} \times \tilde{1})] \qquad (74)$$

$$+ F_A^2 \, [2 \, \text{Re} \, \underline{\mathcal{M}}^* \cdot \hat{\nu} \, \underline{\mathcal{M}} \cdot \tilde{1} + |\underline{\mathcal{M}}|^2 (1 - \hat{\nu} \cdot \tilde{1})$$

$$\left. \mp i \, (\hat{\nu} - \tilde{1}) \cdot (\underline{\mathcal{M}}^* \times \underline{\mathcal{M}})] \right\},$$

where $\hat{\nu} = \underline{\nu}/\nu$, $\tilde{l} = \underline{l}/E_l$, and upper (lower) signs correspond to neutrino (antineutrino) reactions. The nuclear transition matrix elements are

$$\mathcal{M}(\underline{q}) = \langle J_f M_f | \sum_{i=1}^{A} \exp(i\underline{q}\cdot\underline{r}_i)\ \tau_i^{\pm}\ | J_o M_o \rangle \tag{75a}$$

$$\underline{\mathcal{M}}(\underline{q}) = \langle J_f M_f | \sum_{i=1}^{A} \exp(i\underline{q}\cdot r_i)\ \underline{\sigma}_i \tau_i^{\pm}\ | J_o M_o \rangle; \tag{75b}$$

these may be related to similar matrix elements occurring in photoexcitation and electron scattering by isospin rotation using (14). The matrix elements may be written in the form

$$\mathcal{M}(\underline{q}) = \int d^3r\ \exp(i\underline{q}\cdot\underline{r})\ \rho(\underline{r}) \tag{76a}$$

$$\underline{\mathcal{M}}(\underline{q}) = \int d^3r\ \exp(i\underline{q}\cdot\underline{r})\ \rho(\underline{r}), \tag{76b}$$

where the transition densities

$$\rho(\underline{r}) = \langle J_f M_f |\ \rho^{op}(\underline{r}) | J_o M_o \rangle \tag{77a}$$

$$\underline{\rho}(\underline{r}) = \langle J_f M_f |\ \underline{\rho}^{op}(\underline{r}) | J_o M_o \rangle \tag{77b}$$

are the matrix elements of corresponding transition operators for which we perform a multipole expansion /66/

$$\rho^{op}(\underline{r}) = \sum_{i=1}^{A} \delta(\underline{r}-\underline{r}_i)\ \tau_i^{\pm} = \sum_{lm} \tilde{\rho}_{lm}\ Y_{lm}^{*}(\hat{r}) \tag{78a}$$

$$\underline{\rho}^{op}(\underline{r}) = \sum_{i=1} \delta(\underline{r}-\underline{r}_i)\ \underline{\sigma}_i\tau_i^{\pm} = \sum_{ll'm} \tilde{\rho}_{ll'm}(r)\ \underline{Y}_{ll'}^{m*}(r). \tag{78b}$$

Using the Wigner-Eckart theorem, we may express the transition densities as

$$\rho(\underline{r}) = \hat{J}_f^{-1} \sum_{lm} (J_o M_o, lm |\ J_f M_f)\ \rho_l^{of}(r)\ Y_{lm}^{*}(\hat{r}) \tag{79a}$$

$$\underline{\rho}(\underline{r}) = \hat{J}_f^{-1} \sum_{ll'm} (J_o M_o, lm) | J_f M_f)\ \rho_{ll'}^{of}(r)\ \underline{Y}_{ll'}^{m*} \tag{79b}$$

given in terms of their reduced matrix elements

$$\rho_l^{of}(r) \equiv \langle J_f \|\ \tilde{\rho}_l(r) \|\ J_o \rangle, \qquad \rho_{ll'}^{of}(r) \equiv \langle J_f \|\ \tilde{\rho}_{ll'}(r) \|\ J_o \rangle. \tag{80}$$

Inserting (79) into (76) leads to a multipole expansion of the matrix elements

$$\mathcal{M}(\underset{\sim}{q}) = (4\pi/\hat{J}_f) \sum_{LM} (J_oM_o, LM|J_fM_f)\ I_L(q)\ Y^*_{LM}(\hat{q}) \tag{81a}$$

$$\underset{\sim}{\mathcal{M}}(\underset{\sim}{q}) = (4\pi/\hat{J}_f) \sum_{LL'M} (J_oM_o, LM|J_fM_f)\ I_{LL'}(q)\ \underset{\sim}{Y}^{M*}_{LL'}(\hat{q}) \tag{81b}$$

in terms of the reduced transition multipole matrix elements

$$I_L(q) = i^L \int r^2 j_L(qr)\ \rho_L^{of}(r)\ dr \tag{82a}$$

$$I_{LL'}(q) = i^{L'} \int r^2 j_{L'}(qr)\ \rho_{LL'}^{of}(r)\ dr, \tag{82b}$$

which contain the nuclear physics of the problem.

Inserting these results in (74), one finds after some Racah algebra and parity considerations /66/

$$\left(\frac{d\sigma}{d\Omega_1}\right)_{EL} = (1E_1/\pi\hat{J}_o^2) \left\{ F_1^2(1 + \hat{\nu}\cdot\tilde{1})|I_L(q)|^2 \right.$$

$$+ F_A^2 \left[(1 - \tfrac{1}{3}\hat{\nu}\cdot\tilde{1}) + (-)^L\sqrt{6}\ (\hat{\nu}\cdot\hat{q}\ \tilde{1}\cdot\hat{q} - \tfrac{1}{3}\hat{\nu}\cdot\tilde{1})\ \hat{L}^2 \right. \tag{83a}$$

$$\left. \left. \cdot\ (L0,L0|20)\ W(LL11;2L)\right]\ |I_{LL}(q)|^2 \right\}$$

(see also Ref. 13, Appendix A).

$$\left(\frac{d\sigma}{d\Omega_1}\right)_{ML} = (1E_L/\pi\hat{J}_o^2)\ F_A^2 \left[(1 - \tfrac{1}{3}\hat{\nu}\cdot\tilde{1}) \sum'_{L'}|I_{LL'}(q)|^2 \right.$$

$$+ (-)^L\sqrt{6}\ (\hat{\nu}\cdot\hat{q}\ \tilde{1}\cdot\hat{q} - \tfrac{1}{3}\hat{\nu}\cdot\tilde{1}) \sum'_{L'L''} \hat{L}'\hat{L}''\ (L'0,L''0|20) \tag{83b}$$

$$\left. \cdot\ W(L'L''11;2L)\ I^*_{LL'}(q)\ I_{LL''}(q)\right]$$

where $\sum'_{L'}$ means $\sum_{L'=L\pm1}$. Note that here, the vector-axial vector interference and any dependence on neutrino vs antineutrino has disappeared. To retain the latter, one would have to retain terms of order q/2m or relativistic terms. The terminology of "electric" (EL) and "magnetic" (ML) multipoles is the same as for electromagnetic transitions, with parity selection rules

$$\Pi_o \Pi_f = (-)^L \text{ for } I_L \text{ terms (longitudinal or Coulomb),}$$

$$\Pi_o \Pi_f = (-)^L \text{ for } I_{LL} \text{ terms (transverse electric)} \tag{84}$$

$$\Pi_o \Pi_f = (-)^{L+1} \text{ for } I_{LL\pm 1} \text{ terms (transverse magnetic)}$$

(Coulomb and transverse electric transitions have the same selection rule and are given the common designation EL).

4.3 Collective Nuclear Excitations Described by the Goldhaber-Teller Model

Excitation of nuclear giant dipole states by neutrinos was first considered by BELYAEV /126/, who related the probability of this reaction to the photonuclear giant dipole cross section. This approach is equivalent to that used later on by FOLDY and WALECKA, in the case of muon capture, cf. Sect. 3.1. In principle, if one considers nuclei whose ground-state may be approximated by scalar multiplets, both vector and axial vector matrix elements can be obtained in this way if SU4 symmetry is invoked /31/, although the axial vector was disregarded by BELYAEV. Another convenient method for the calculation of both vector and axial neutrino excitations of giant resonance levels is provided by the use of the Goldhaber-Teller model in its generalized form /51/ including spin-isospin vibrations (cf. Sect. 1.4). This model, which was discussed in Sect. 1.4, gives after isospin rotation, i.e., a replacement of the operators

$$\rho_i^{op}(\underline{r}) \equiv \frac{1}{2} \sum_{i=1}^{A} \delta(\underline{r}-\underline{r}_i)\, \tau_{3i} \rightarrow \mp \frac{1}{\sqrt{2}} \sum_{i=1}^{A} \delta(\underline{r}-\underline{r}_i)\, \tau_i^{\pm} \tag{85}$$

see (22a), the expressions for the transition densities of charge-exchange dipole transitions in self-conjugate ($T_0 = 0$) nuclei

$$\begin{aligned} \rho_i(\underline{r}) &= \pm \frac{1}{\sqrt{2}}\, \underline{d} \cdot \underline{\nabla}\, \rho_0(r) \\ \rho_{si}(\underline{r}) &= \pm \frac{1}{\sqrt{2}}\, \delta_{im'}^{(s)}\, \underline{d} \cdot \underline{\nabla}\, \rho_0(r), \end{aligned} \tag{86}$$

where $\underline{d}$ is the displacement between the two nucleon spheres in the model (see Sect. 1.4), and $\rho_0(r)$ the (spherically symmetric) nuclear ground-state density; we also use the spherical Kronecker tensor

$$\delta_{im'}^{(S)} = \begin{cases} -2^{-1/2}(\delta_{i1}+i\delta_{i2}) & m' = 1 \\ \delta_{i3} & m' = 0 \\ 2^{-1/2}(\delta_{i1}-i\delta_{i2}) & m' = -1 \end{cases} \tag{87}$$

m' being the spin projections of the total spin S = 1 of the s-i vibration. Inserting (86) into (76) [using $\underline{d}$ (19)], we obtain

$$\mathcal{M}(\underline{q}) = iqF(q)\left(\frac{2\pi}{3Am\omega}\right)^{1/2} Y^*_{1M}(\hat{q}) \tag{88a}$$

$$\mathcal{M}_i(\underline{q}) = \delta_{im'}^{(S)}\ \mathcal{M}(q), \tag{88b}$$

where the final spin projections M of L ≡ 1 and m' of S = 1 in (88) require coupling according to the final-state nuclear spin $\underline{J}_f = \underline{L} + \underline{S}$. Here, F(q) is the nuclear ground state form factor

$$F(q) = \int d^3r\ \rho_0(r)\ e^{i\underline{q}\cdot\underline{r}} \tag{89}$$

normalized to Z at q = 0.

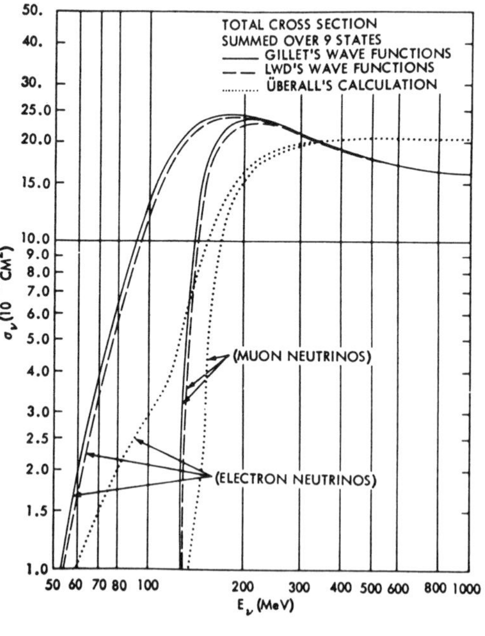

Fig. 13. Neutrino excitation cross section corresponding to giant dipole transitions /128/ calculated with the Goldhaber-Teller model (dotted curves) and with shell-model wave functions (see Sect. 4.5) for reaction of (72) for l = e or μ on a ^{12}C target

A calculation of the charged-current cross section of giant dipole excitations for both isospin and spin-isospin transitions in ^{12}C was performed by ÜBERALL /51/ using the generalized Goldhaber-Teller model, with results shown in Figure 13 /128/ (dotted curves) for both $l = e$ and $l = \mu$. Figure 14 shows similar results obtained by DONNELLY et al. /129/ for (ν_e, e^-) (solid lines) and $(\bar{\nu}_e, e^+)$ (dashed lines) reactions on ^{16}O and ^{40}Ca, for the spin-flip giant dipole transitions which provide the dominant contributions. A difference between ν and $\bar{\nu}$ cross sections appears here, due to the retention in this work of terms of order q/2m in the interaction Hamiltonian.

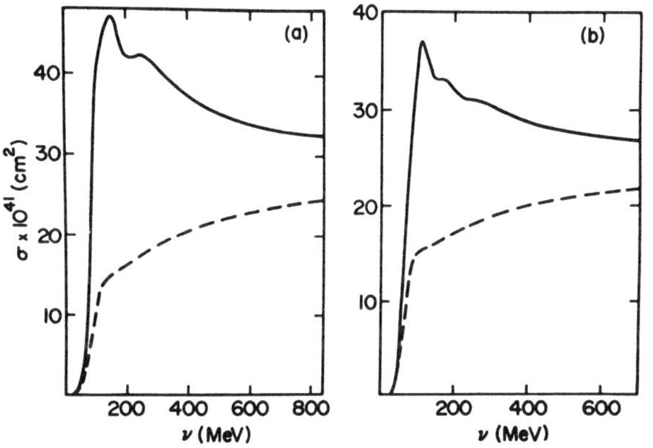

Fig. 14. Neutrino (solid curve) and antineutrino (dashed curve) spin-isospin giant dipole cross sections in a) ^{16}O and b) ^{40}Ca calculated using the Goldhaber-Teller model /129/ plotted vs neutrino energy. Note that the ^{40}Ca excitation cross section is smaller than in ^{16}O due to retardation effects (form factor)

The neutrino excitation of nuclei via neutral currents, i.e., via the reaction

$$\nu_l + A_Z \rightarrow A_Z^{(*)} + \nu_l \tag{90}$$

was first discussed by KING /130/ and also by GERSHTEIN et al. /131/. Calculations of giant resonance excitations by the reaction of (90), using the Goldhaber-Teller model, were carried out by DONNELLY et al. /129/ and by DADAYAN /132/ (see also /133/). Figure 15 shows the results of /129/ for ^{16}O and ^{40}Ca.

An experimental test for the effects of neutral current interactions involving giant resonance excitations in nuclei is now being carried out at the LAMPF neutrino facility /134/ via the reactions

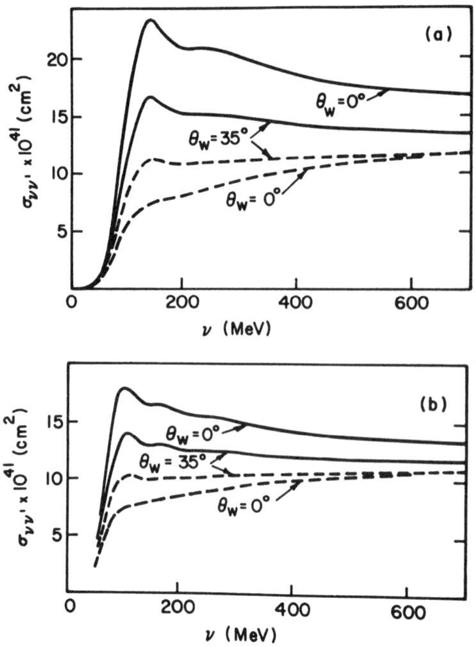

Fig. 15. Neutrino (solid curve) and antineutrino (dashed curve) giant dipole cross sections in a) ^{16}O and b) ^{40}Ca calculated using the Goldhaber-Teller model /129/, for Weinberg angles θ_W = 0° and 35°, plotted vs neutrino energy, corresponding to the Eq. (90)

$$\nu_\mu + {}^{39}K \rightarrow \nu_\mu + {}^{37}Ar + {}^{2}H \text{ (or N + P)} \tag{91a}$$

and /135/

$$\nu_\mu + {}^{40}Ca \rightarrow \nu_\mu + {}^{37}Ar + {}^{3}He \text{ (or N + 2P)} \tag{91b}$$

using ν_μ's that result form the decay of pions in flight. The reactions are observed by detecting the activity of ^{37}Ar (35-day half-life) via its 2.8 keV Auger electron.

4.4 Sum Rules for Forward Neutrino-Induced Reactions

Using sum rule techniques, the forward neutrino scattering cross section may be evaluated in terms of ground-state expectation values, similarly as in Sect. 2.2, where procedures of this kind were used for electron scattering. Giant dipole resonances are predominantly excited in the forward scattering direction where q is relatively small, rather than higher spin states which require larger q for their effective excitation.

In the forward direction, kinematics gives for high lepton energies so that $E_l \gg m_l$

$$q \cong \nu - E_l = \omega. \tag{92}$$

Accordingly, one has for the 4-momentum transfer $q_\lambda^2 \cong 0$, and one may thus use the weak currents of (11) in which no q_λ^2 dependence of the form factors appears. Moreover, the terms in (11) containing $\sum_i \underline{\sigma}_j \cdot \underline{\nabla}$ are suppressed if acting on nuclear ground-states that are approximately scalar supermultiplets (as for ^4He, ^{16}O, ^{40}Ca), cf. [Ref. 34, Sect. 3]. With these assumptions, i.e., dipole dominance of the non-relativistic operator, relativistic terms $\sim v_i$, and closure approximation, we obtain apart from kinematical factors /136/

$$\begin{aligned}\left(\frac{d\sigma}{d\Omega}\right)_{forward} &= - <0|\ \textstyle\sum_i g_V(i\underline{q}\cdot\underline{r}_i)\ \tau_i^-\ \sum_j g_V(i\underline{q}\cdot\underline{r}_j)\ \tau_j^+\ |0> \\ &- <0|\ \textstyle\sum_i g_A(\underline{\sigma}_i\cdot\underline{\sigma})(i\underline{q}\cdot\underline{r}_i)\ \tau_i^-\ \sum_j g_A(\underline{\sigma}_j\cdot\underline{\sigma})(i\underline{q}\cdot\underline{r}_j)\ \tau_j^+\ |0> \\ &- <0|\ \left[\textstyle\sum_i g_V(i\underline{q}\cdot\underline{r}_i)\ \tau_i^-,\ \sum_j g_V(\underline{\sigma}\cdot\underline{\nu})(\underline{\sigma}\cdot\underline{v}_j)\ \tau_j^+\right] |0> \\ &+ <0|\ \left[\textstyle\sum_i g_A(\underline{\sigma}\cdot\underline{\sigma}_i)(i\underline{q}\cdot\underline{r}_i)\ \tau_i^-,\ \sum_j g_A(\underline{\sigma}_j\cdot\underline{v}_j)\ \tau_j^+\right] |0> \\ &+ <0|\ \textstyle\sum_i g_A(\underline{\sigma}\cdot\underline{\nu})(\underline{\sigma}_i\cdot\underline{v}_i)\ \tau_i^-\ \sum_j g_A(\underline{\sigma}\cdot\underline{\nu})(\underline{\sigma}_j\cdot\underline{v}_j)\ \tau_j^+\ |0> \\ &- <0|\ \textstyle\sum_i g_V(\underline{\sigma}\cdot\underline{\nu})(\underline{\sigma}\cdot\underline{v}_j)\ \tau_i^-\ \sum_j g_V(\underline{\sigma}\cdot\underline{\nu})(\underline{\sigma}\cdot\underline{v}_j)\ \tau_j^+\ |0>.\end{aligned} \tag{93}$$

These expressions may be evaluated by double-commutator sum rules as (37a,b), with the following results:

The first term is the usual dipole operator, giving

$$g_V^2\ \omega_R(A/2m). \tag{94a}$$

Its contribution is canceled by that of the relativistic third and sixth terms, in agreement with ADLER's theorem /137/ which states that when the lepton in $\nu + \alpha \to l + \beta$ emerges in the forward direction with an energy $E_l \gg m_l$ and $m_\alpha \neq m_\beta$, the squared matrix element depends only on the divergences of the vector and axial currents, and the vector divergence is then eliminated by the conserved vector current (CVC) hypothesis. This illustrates the importance of retaining the weak vector current contributions, (11d), in addition to those of the charge, (11e), [which, e.g., was solely retained in (71b)] for forward neutrino reactions. Note that such a cancellation would not occur for allowed Fermi transitions between members of an isospin multiplet, induced by $\rho_{vec}^{nuc\pm}$; but this does not contradict ADLER's theorem because $m_\alpha = m_\beta$ for the multiplet.

The remaining terms furnish the result

$$g_A^2(\omega_{Rs}/m)\left[A - \frac{4}{3} \langle 0| \sum_i \underline{l}_i \cdot \underline{s}_i |0\rangle\right], \qquad (94b)$$

where ω_{Rs} is the energy of the spin-flip resonance; but in our approximation of the nuclear Hamiltonian containing Wigner forces only, and the nuclear ground-states being scalar supermultiplets, the $\underline{l} \cdot \underline{s}$ term in (94b) will not contribute. For these axial terms, the contribution of the relativistic component (11c) is found to be as important as the (first-forbidden) Gamow-Teller giant dipole spin-flip term.

In the case of nuclei containing ground-state impurities regarding SU4 symmetry (such as ^{12}C), corrections would have to be added both in the derivation of the sum rules, and with respect to the retention of interaction terms inducing allowed transitions, in particular Gamow-Teller terms that lead to analog M1 transitions. In ^{12}C, these were evaluated both by the shell model /138/ and the Helm model /66/, and were found to be dominant in the forward direction. In ^{16}O, a 1^+ transition exists also which has a small electroexcitation strength but becomes somewhat more important in forward neutrino scattering /139/.

4.5 Helm Model and Shell-Model Calculations

The generalized Helm model /140/ is a convenient phenomenological nuclear model which can be used for a rapid parameterization of the transition matrix elements of nuclear reactions. Since similar matrix elements enter in the cross section expressions of a large variety of intermediate-energy reactions (see Sect. 1), a procedure offers itself to obtain the model parameters from a fit of theoretical form factors to the experimental data of a few especially well-measured processes (usually, inelastic electron scattering), and subsequently to utilize the same model with the now known parameters to predict or interpret the experimental results of other reactions. This type of phenomenological application is to be considered the principal merit of the Helm model, since it cannot provide any detailed understanding of the nuclear structure.

The model considers the reaction to take place in the vicinity of the nuclear radius, which is justified in view of the actual overlap of initial and final nuclear wave functions in this region. Transition densities are therefore taken to have a delta-function peak at a transition radius R, while a Gaussian folding of the transition operators (which simulates the width of the transition region) is performed in the fashion

$$\rho^{op}(\underline{r}) = \int \rho_R^{op}(\underline{r}') \, \rho_g(\underline{r}-\underline{r}') \, d^3r' \tag{95a}$$

with

$$\rho_g(\underline{r}) = (2\pi g^2)^{-3/2} \exp(-r^2/2g^2). \tag{95b}$$

The "intrinsic" transition operator ρ_R^{op} (and similarly $\underline{J}_{\overline{R}}^{op}$) are multipole expanded as in (78), and the reduced matrix elements $J_{1 1'}^{R}(r)$, $\rho_1^{\overline{R}}(r)$ are introduced as in (80). Writing

$$\underline{J}_{em}(\underline{r}) = \underline{j}_{em}(\underline{r}) + \underline{\nabla} \times \underline{\mu}\,(\underline{r}), \tag{96a}$$

i.e., separating the current into a convection (j_{em}) and a magnetization current due to a magnetization density (μ),

$$\underline{\mu}^{op}(\underline{r}) = \frac{1}{2m} \sum_{j=1}^{A} \delta(\underline{r}_j-\underline{r}) \left[\tfrac{1}{2}(\mu_p+\mu_n) + \tfrac{1}{2}(\mu_p-\mu_n)\,\tau_{3j} \right] \underline{\sigma}_j, \tag{96b}$$

we now take the intrinsic densities to be delta functions so that

$$\rho_1^R(r) = i^{-L}\hat{J}_i(\beta_1/R^2)\,\delta(r-R) \tag{97a}$$

$$j_{1 1+1}^R(r) = 0 \tag{97b}$$

$$j_{1 1}^R(r) - i^{-L}\hat{J}_i\beta_{11}(4\pi/3Z)\,\rho_R(r)\,(r/R)^1 \tag{97c}$$

$$j_{1 1-1}^R(r) = -i^{-L}\hat{J}_i(4\pi/3Z)\,(i\omega)\,R\rho_R(r)\beta_1(\hat{1}/1^{1/2})\,(r/R)^{1-1} \tag{97d}$$

$$\mu_{1 1'}^{\overline{R}}(r) = i^{-L'}\hat{J}_i\gamma_{11'}(1/2m\overline{R}^2)\,\delta(r-\overline{R}). \tag{97e}$$

The form of $j_{1 1-1}^R$ arises from satisfying the continuity equation (6), and that of $j_{1 1+1}$ from the assumed nuclear incompressibility /141/. A magnetic convection current $j_{1 1}^R$ is introduced, but in general its contribution is small so that $\beta_{11} \cong 0$. Further, ρ_R is the step function density

$$\rho_R(r) = 3Z/4\pi R^3 \;(r<R),\; 0\,(r>R). \tag{97f}$$

With a multipole expansion of the electromagnetic transition densities of $\rho(\underline{r})$, $\underline{j}(\underline{r})$, $\underline{\mu}(\underline{r})$ as in (79), inserted in (3a), (3b), and (9), and using (96a), one finds for the electron scattering form factors /13/

$$\mathcal{M}_L(q) = i^L \int r^L j_L(qr)\, \rho_L(r)\, dr \tag{98a}$$

$$\begin{aligned}\mathcal{T}_L^{el}(q) = & -i^{L-1} \left[(L+1)^{1/2}/\hat{L}\right] r^2 j_{L-1}(qr)\, j_{LL-1}(r)\, dr \\ & -i^{L+1}(L^{1/2}/\hat{L}) \int r^2 j_{L+1}(qr) j_{LL+1}(r)\, dr + i^L q \int r^2 j_L(qr)\, \mu_{LL}(r)\, dr\end{aligned} \tag{98b}$$

$$\begin{aligned}\mathcal{T}_L^{mag}(q) = & -i^{L-1} q \left[(L+1)^{1/2}/\hat{L}\right] \int r^2 j_{L-1}(qr)\, \mu_{LL-1}(r)\, dr \\ & -i^{L+1} q\, (L^{1/2}/\hat{L}) \int r^2 j_{L+1}(qr)\, \mu_{LL+1}(r)\, dr \\ & +i^L \int r^2 j_L(qr) j_{LL}(r)\, dr.\end{aligned} \tag{98c}$$

If tne intrinsic densities are multipole expanded (as appropriate for the Helm model) and the operators convoluted as in (95a), then (98) would contain $\rho_L^R(r)$, etc., and would have as overall factors the Fourier transform of (95b), i.e.,

$$f_g(q) = e^{-g^2 q^2/2}. \tag{99}$$

Inserting (97) one thus obtains the Helm model form factors for electron scattering,

$$\mathcal{M}_L(q) = \hat{J}_i\, \beta_L\, f_g\, (q)\, j_L(qR) \tag{100a}$$

$$\begin{aligned}\mathcal{T}_L^{el}(q) = \hat{J}_i \Big\{ & \left[(L+1)L\right]^{1/2} \beta_L(\omega/q)\, f_g(q)\, j_L(qR) \\ & + \gamma_{LL}(q/2m)\, f_{\bar{g}}(q)\, j_L(q\bar{R}) \Big\}\end{aligned} \tag{100b}$$

$$\begin{aligned}\mathcal{T}_L^{mag}(q) = & -\hat{J}_i(q/2m)\, f_{\bar{g}}(q) \Big\{ (L^{1/2}/\hat{L})\, \gamma_{LL+1}\, j_{L+1}(q\bar{R}) \\ & + \left[(L+1)^{1/2}/\hat{L}\right] \gamma_{LL-1}\, j_{L-1}(q\bar{R}) \Big\} + \hat{J}_i(\beta_{LL}/qR)\, f_g(q)\, j_{L+1}(qR).\end{aligned} \tag{100c}$$

They depend on the radial and surface smearing parameters R, g ($\bar{R}, \bar{g}$) of charge-current (magnetization) densities, and on the corresponding strength parameters β_L, β_{LL} (γ_{LL}, $\gamma_{LL\pm1}$), respectively. A fit, using this del, to the longitudinal and transverse giant resonance form factors of electron scattering from ^{12}C is shown in Figure 4, which fixes the Helm parameters for this collective transition. A Helm model fit to the form factor of the 15.11 MeV, 1^+ level (giant M1 resonance) in ^{12}C is presented in Figure 16.

In this way, the Helm model parameters of the T = 1 levels observed in ^{12}C and ^{16}O by electron scattering have been determined (see /66/, /142/ for ^{12}C, and /139, 143, 144/ for ^{16}O); they are reproduced in Tables 5 and 6. For ^{12}C, we show the corresponding level scheme in Figure 17, together with the analog levels of the neighboring nuclei reached by charge-exchange reactions.

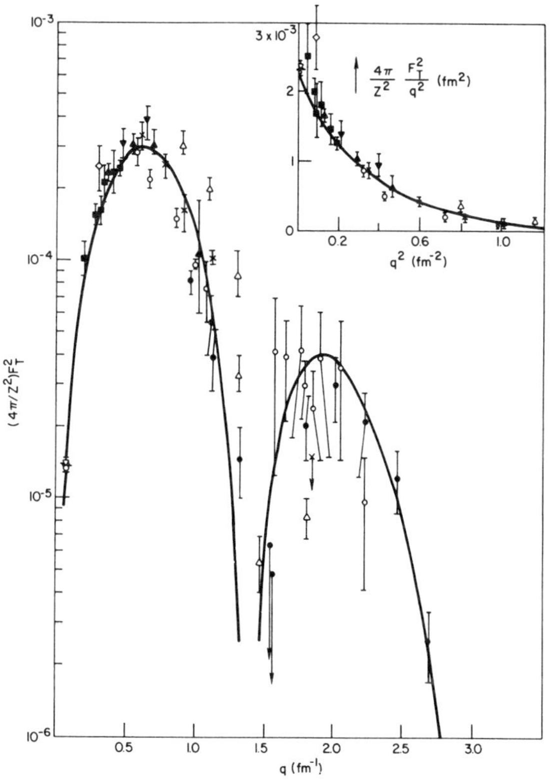

Fig. 16. Squared form factor of the 15.11, 1^+ level of ^{12}C vs q, fit by the Helm model with parameters given in Table 5

Table 5. Helm model parameter of T = 1 levels in ^{12}C /66, 142/. Here ω is the level excitation energy in ^{12}C. Two versions of the giant resonance are presented (No. 18, 19), integrated over two different energy regions as indicated. Further, some individual peaks inside the giant resonance region (No. 14-17) are presented separately. Levels No. 6 and 12 were observed in electroexcitation of ^{12}C but have no counterpart in ^{12}B so that their assignment of T = 1 is questionable

Level No.	ω (MeV)	J^{π}	σL	g [fm]	R [fm]	β_L	$\bar{g}$ [fm]	$\bar{R}$ [fm]	γ_{LL}	γ_{LL-1}	γ_{LL+1}
1	15.11	1^+	M1	-	-	-	0.90	1.92	-	0.89	-0.78
2	16.11	2^+	E2	0.69	2.61	0.173	0.69	2.24	1.14	-	-
3	16.58	2^-	M2	-	-	-	0.83	2.58	-	0.69	-2.64
4	17.23	1^-	E1	0.65	2.00	0.091	0.65	1.85	0.51	-	-
5	17.77	0^+	C0	-	-	-	-	-	-	-	-
6	18.15	1^-	E1	-	-	0	0.77	3.47	0.812	-	-
7	18.72	3^-	E3	0.77	2.85	0.315	-	-	0	-	-
8	18.81	2^+	E2	0.77	2.67	0.037	0.77	2.67	0.019	-	-
9	19.2	1^-	E1	0.77	2.08	0.048	0.77	2.08	0.338	-	-
10	19.4	2^-	M2	-	-	-	0.77	2.97	-	1.47	-0.99
11	19.6	4^-	M4	-	-	-	0.77	2.82	-	2.68	2.59
12	20.0	2^+	E2	0.77	2.50	0.123	0.77	2.50	0.421	-	-
13	20.6	3^+	M3	-	-	-	0.77	2.40	-	0.72	2.22
14	21.6	3^-	E3	0.77	2.82	0.243	0.77	2.82	0.700	-	-
15	22.0	1^-	E1	0.77	2.50	0.253	-	-	0	-	-
16	22.7	1^-	E1	-	-	0	0.77	2.60	0.829	-	-
17	23.8	1^-	E1	0.77	2.50	0.207	-	-	0	-	-
18	21-26	1^-	E1	0.77	2.08	0.438	0.77	2.08	2.230	-	-
19	21-37	1^-	E1	0.77	2.08	0.720	0.77	2.08	3.860	-	-
20	24.9	0^-	E1s	-	-	-	-	-	-	-	-
21	25.5	3^-	E3	-	-	-	-	-	-	-	-
22	34.0	0^-	E1	-	-	-	-	-	-	-	-

Table 6. Helm model parameters of T = 1 levels in ^{16}O. Here, ω is the level excitation energy in ^{16}O, and $\bar{g}$ = g throughout. The parameter δ_L ($0 \leqq |\delta_L| \leqq 1$) represents an empirical factor which for the 1^- (and 3^-) giant resonance levels multiplies the β_L γ_{LL} cross terms of the squared electron scattering form factors, expressing partial coherence between the charge-current and magnetization transitions of these levels caused by spin-orbit coupling, see (100b) /139, 143, 144/

Level No.	ω (MeV)	J^π	σL	g [fm]	R [fm]	β_L	$\bar{R}$ [fm]	γ_{LL}	γ_{LL-1}	γ_{LL+1}	δL
2	12.967	2^-	M2	0.771	-	-	2.65	-	-1.243	0.313	-
3	13.093	1^-	E1	0.782	3.44	-0.0604	3.40	0.485	-	-	-
4	13.258	3^-	E3	0.577	2.37	0.539	3.36	0.710	-	-	-
5	16.21	1^+	M1	0.672	-	-	3.22	-	0.325	-0.160	-
6	16.79	3^+	M3	0.942	-	-	2.40	-	2.265	-1.940	-
7	17.30	1^-	E1	0.520	2.60	0.0689	3.39	0.401	-	-	-
8	17.60	2^-	M2	0.520	-	-	2.26	-	0.480	1.090	-
9	18.50	2^+	E2	0.482	3.14	0.214	2.22	1.11	-	-	-
10	18.99	1^-	E1	0.593	2.61	0.0579	2.75	-1.067	-	-	-
11	19.04	2^-	M2	1.09	-	-	3.90	-	1.530	-4.350	-
12	19.04	4^-	M4	0.723	-	-	3.21	-	3.820	1.640	-
13	19.48	1^-	E1	0.613	2.61	0.0552	2.75	-1.058	-	-	-
14	20.36	2^-	M2	0.920	-	-	2.31	-	-2.60	4.24	-
15	21.02	1^-	E1	0.920	3.41	-0.070	4.23	0.998	-	-	-
16	23	1^-	E1	0.824	2.83	0.474	2.35	0.747	-	-	0.217
17	23	2^-	M2	0.928	-	-	3.12	-	3.57	2.32	-
18	23	3^-	E3	0.548	3.06	1.21	2.60	2.539	-	-	-0.459

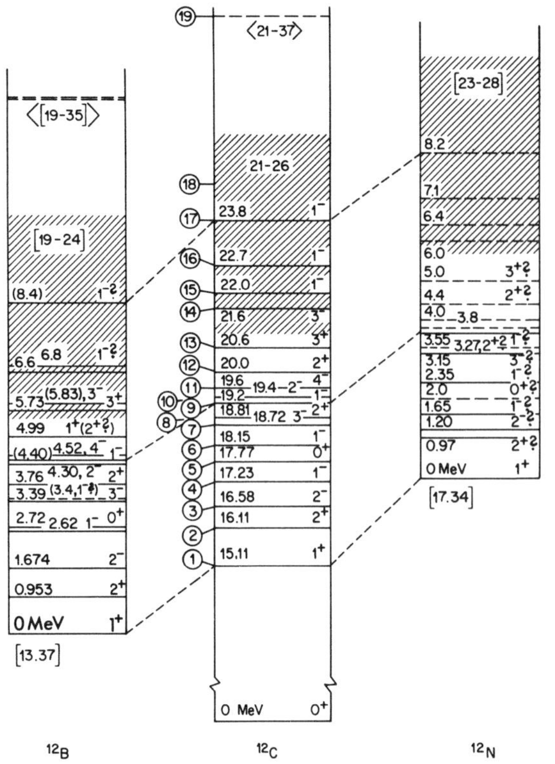

Fig. 17. T = 1 levels of ^{12}C as observed in electron scattering, and analog levels in the neighboring nuclei reached by charge-exchange reactions

For the neutrino reactions, the corresponding transition densities (79), are obtained from the preceding ones of electron scattering by isospin rotation, cf. (85). This means essentially that $\mp \frac{\sqrt{2}}{e}\,\rho(r)$ of electron scattering represents $\rho(r)$ of the charge-exchange neutrino reaction. Further, $\rho_{11'}(r)$ of (79, 80) is related to the magnetization density of electron scattering by

$$\rho_{11'}(r) = \mp\sqrt{2}\,\frac{2m}{e(\mu_p - \mu_n)}\,\mu_{11'}(r). \tag{101}$$

The Helm model expressions for (82) are therefore

$$I_L(q) = \frac{\sqrt{2}}{e}\,\hat{J}_i \beta_L f_g(q)\, j_L(qR) \tag{102a}$$

$$I_{LL'}(q) = \sqrt{2}\,\hat{J}_i\,\frac{\gamma_{LL'}}{e(\mu_p - \mu_n)}\, f_{\bar{g}}(q)\, j_{L'}(q\bar{R}). \tag{102b}$$

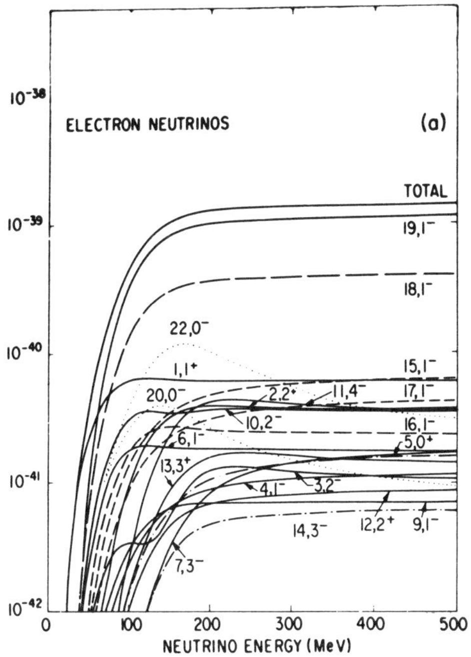

Fig. 18. Electron neutrino cross sections for excitation of the ^{12}N levels as shown in Figure 17, calculated using the Helm model /66/

The corresponding cross sections for neutrino reactions (72) exciting the T = 1 analog levels from ^{12}C and ^{16}O targets were calculated in /66/ and /139/, respectively, using the Helm model. The results for total ^{12}C cross sections are shown in Figure 18 for electron neutrinos. It is seen that while at small energies the contribution of the 1^+ level dominates, the giant resonance is dominant at high energies (in this figure, levels No. 1-4 were calculated with parameters given in /66/, which differ slightly from those given in Table 5).

Several shell-model calculations of charge-exchange neutrino reactions (72) in ^{12}C have also been carried out, using a simple particle-hole model with configuration mixing for a description of the giant resonances /128, 145, 146/. The results of one of these calculations /128/ is shown in Figure 13, using

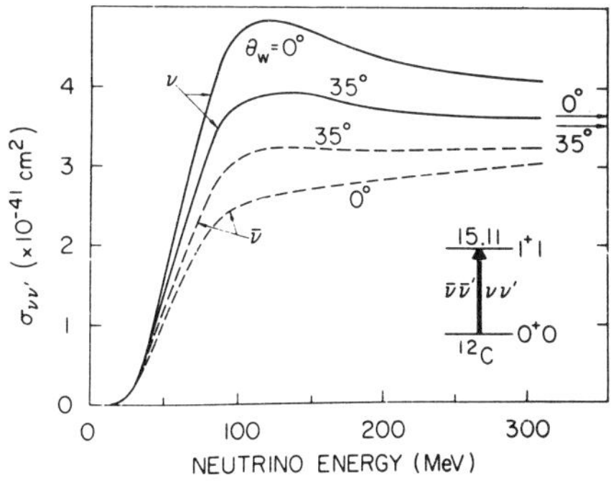

Fig. 19. Neutral-current inelastic neutrino (solid curves) and antineutrino (dashed curves) scattering cross sections to the 1^+, T = 1 (15.11 MeV) state in ^{12}C, for two values of the Weinberg angle θ_W as indicated /149/

particle-hole wave functions of LEWIS, WALECKA, and DeFOREST (LWD) /71, 147/, or of GILLET /148/. Shell-model calculations of neutral-current reactions (90) were also performed for T = 0 and T = 1 levels in ^{12}C and ^{16}O, and we show some of these results for the excitation of the 15.11, 1^+ level in ^{12}C /149/ in Figure 19, and of negative-parity T = 1 levels of ^{16}O (with shell-model energies) in Figure 20 /150/. These reactions, especially to low-lying T = 0 levels, may prove useful for a detection of neutrino bursts originating from the gravitational collapse of stars, where the neutrino spectrum peaks below ~10 MeV /151/ so that charge-exchange processes cannot take place in light nuclei.

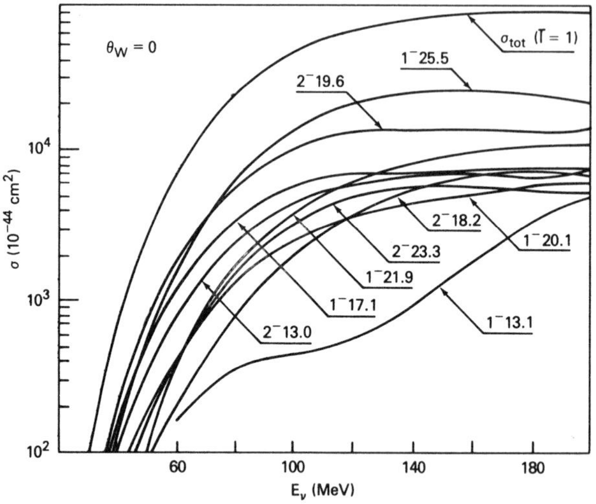

Fig. 20. Shell-model calculation of neutral-current neutrino cross sections exciting T = 1 negative-parity levels in ^{16}O, with Weinberg angle $\theta_W = 0$ /150/

5. Photoproduction and Radiative Capture of Pions

5.1 Giant Resonance Excitation in Reactions Involving Pions

In the preceding sections, we discussed semileptonic weak and electromagnetic interactions, whose quantitative calculations are quite reliable to first order because of the smallness of their coupling constants, and which in nuclear matter may also be described reasonably well in impulse approximation since they correspond to (weak) contact interactions (except for the pseudoscalar term), or (electron-scattering) one-photon exchange terms which are little affected by nuclear matter. In this and the following section, however, we turn to reactions that involve hadrons, whose elementary coupling with the nucleons is in general "strong" and in whose description two-step processes can enter that involve nuclear or isobar propagators. The latter may be affected in a nuclear environment, so that a use of the impulse approximation is less reliable. In the case of charged pion photoproduction on a nucleus,

$$\gamma + A_Z \to A^{(*)}_{Z\mp 1} + \pi^{\pm} \qquad (103)$$

for example (to mention just the Born terms), the contact term coming from the low energy expansion of the Feynman diagrams of Figure 21a which is determined by the

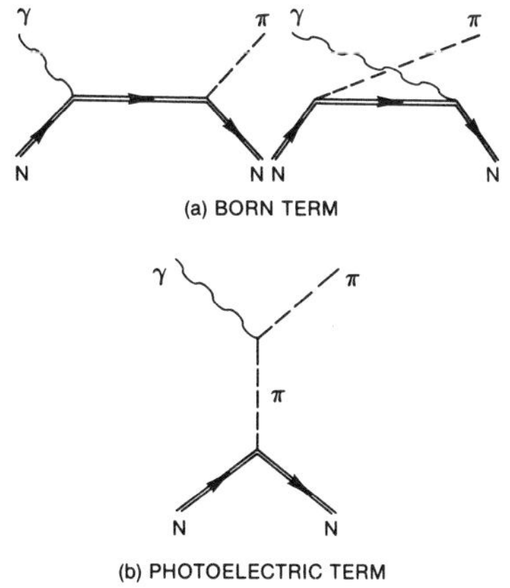

Fig. 21. Born terms in charged pion photoproduction

KROLL-RUDERMAN theorem /152/, is supplemented by the "photoelectric" term, Figure 21b. The latter, similar to the pseudoscalar term in the weak interactions, is of comparatively long range, and the pion propagator is expected to be affected in the nuclear medium, as is the (shorter range) nucleon propagator in Figure 21.

If treated by an appropriate t-matrix approach together with the use of impulse approximation, the matrix elements describing these hadronic processes are found to be of similar structure as those that enter in the weak and electromagnetic processes. It is this fact that permits the treatment of electromagnetic, weak, and hadronic intermediate-energy reactions via a unified approach, as discussed in the first paragraph of Sect. 4.5.

In the present section, photopion reactions will be considered which can be characterized by an effective scattering length $a_{\pi N\gamma} \sim 10^{-2}$fm, so that their quantitative description (at least at threshold) should by fairly reliable.

Depending on the type of reaction considered, the transition amplitude will exhibit a different spin and isospin structure, so that as far as giant resonance is concerned, its various spin or isospin components can be selectively excited by different reactions. At threshold, the photopion S-wave transition amplitude has the form (for the production of $\pi^{\pm}$) /152/

$$a_{\pi N\gamma} = i\, a_0^{\pm}\, \tau^{\mp}\, \underline{\sigma}\cdot\underline{\varepsilon} \tag{104a}$$

(contact term), where $\underline{\tau}$ and $\underline{\sigma}$ are the isospin and spin operators of the nucleon, $\underline{\varepsilon}$ is the photon polarization vector, and $a_0^{\pm}$ ($\sim 10^{-2}$fm) independent of τ and σ [cf. (106)]. This shows that S-wave pion photoproduction and radiative pion capture will selectively excite /153/ the spin-isospin waves, out of the four possible types of collective oscillations [density waves (c), spin waves (s), isospin waves (i), and spin-isospin waves (si) - apart from all possible orbital multipolarities - as discussed in Sect. 1.4]. Similarly, the scattering of slow pions, which is goverened by the S-wave transition amplitude (of the order $\sim 10^{-1}$fm) /24/,

$$a_{\pi N} = b_0 + b_1\, \underline{\tau}\cdot\underline{t}_{\pi}, \tag{104b}$$

(where $\underline{t}_{\pi}$ is the pion isospin operator, and b_0 and b_1 are constants that do not depend on $\underline{\tau}$ or $\underline{t}_{\pi}$) will selectively excite density (c) and isospin (i) waves. With increasing energy, however, photopion reactions will begin to excite also isospin collective oscillations in addition to the si waves, and pion scattering will begin to excite s and si oscillations, in addition to the c and i waves.

Of the two types of photopion reactions, i.e., photopion production (103), and its inverse, namely the capture of stopped pions from a bound atomic orbit,

$$\pi^- + A_Z \rightarrow A^{*}_{Z-1} + \gamma \tag{105a}$$

the former has the same advantage over the latter which electron scattering has over photon absorption, namely a variable range of momentum transfer q. This permits a study of the nuclear form factors as functions of q, rather than their determination at a fixed value of q. Radiative pion capture in flight,

$$\pi^{\pm} + A_Z \rightarrow A_{Z\pm 1}^{(*)} + \gamma \qquad (105b)$$

also has the advantage of a variable q /154/; although these experiments are hard, they appear to have become possible now /155/.

Pion charge exchange scattering ($\pi^{\pm}$, π^{0}) on nuclei /156/ may constitute a nuclear probe which is potentially competitive with (p,n) and (n,p) processes /157/ in studying isobaric analog resonances (see also Sect. 4.4).

5.2 Radiative Pion Capture

A number of review articles exist on this subject /39, 41, 158-160/, with a considerable amount of experimental material now becoming available /160/. At the same time, some detailed calculations are being carried out for both light /121, 161/ and heavy nuclei /162/, so that the field is developing rapidly at this time.

We shall here concentrate on the process of (105a), in which a negative pion is slowed down in matter and captured in a bound atomic orbit. This process requires an atomic cascade calculation for its interpretation /41/, in order to ascertain the initial atomic orbit from which the pion is captured by the nucleus. For ^{16}O, e.g., the pion, due to its strong interaction with the nucleons, is captured about 91% of the time while still in the p atomic orbit, and in only 9% of the time reaches the s orbit from where it is captured then. In some light nuclei, however (such as ^{6}Li), it is experimentally possible to study the captures that took place

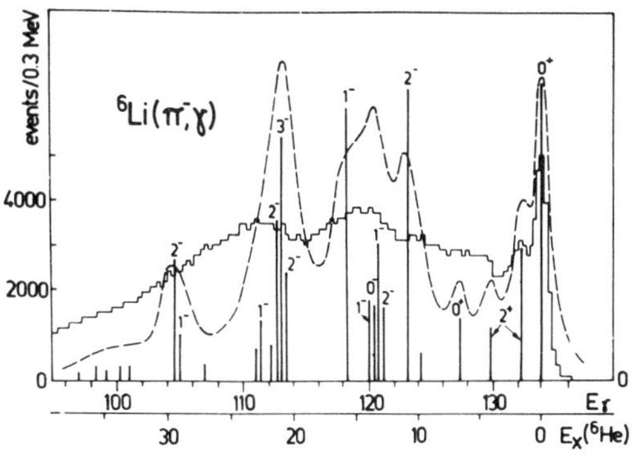

Fig. 22. Photon spectrum following 1s-orbital radiative pion capture in ^{6}Li /161/

from the 1s-orbit exclusively, by observing the coincident 2p→1s x-ray /163/, so that the comparison with theory is more direct here. This is illustrated in Figure 22, which shows the experimental photon spectrum following 1s-orbital capture (π^-, γ) in ^{6}Li, obtained by reference /163/ as a histogram. This is compared with a theoretical spectrum (solid lines), spread over Breit-Wigner shapes (dashed curves), obtained using a configuration-mixed shell model by KISSENER et al. /161/ which includes the description of the analog giant resonance. The lower scale of the figure shows the energy of the final states in ^{6}He.

The figure illustrates one of the possible experimental methods to measure the radiative capture of stopped negative pions, namely by an observation of the high-energy photon spectrum. This method was pioneered by BISTIRLICH et al. /164/ using an electron-positron pair spectrometer for this purpose. The early calculations for the process /165, 166/, based on Goldhaber-Teller model or shell-model assumptions, obtained both the photon spectra, and the spectra of neutrons emitted in the subsequent decay of the giant resonance excited during the capture.

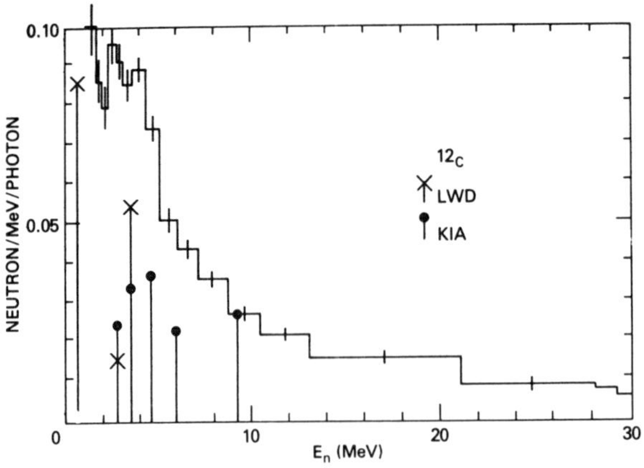

Fig. 23. Integrated neutron spectrum following radiative pion capture in ^{12}C /167/

Observation of these decay neutrons in coincidence with the emitted photon (which permits a measurement of the neutron spectrum by the time-of-flight method) was achieved by LAM and collaborators /167/. An example of neutron spectra measured in this way from radiative pion capture in ^{12}C is shown in Figure 23, together with theoretical shell-model results (solid lines) /166/ based on wave functions of LEWIS, WALECKA, and DeFOREST (LWD) /71, 147/, and of KAMIMURA, IKEDA, and ARIMA (KIA) /168/.

The larger the Z of the nucleus, the closer the bound atomic orbits of the π^- will be to the nuclear surface. Therefore, the capture occurs predominantly from the 1s-orbit only for the very lightest nuclei (^2H, ^4He), and from the p-orbit for other light nuclei (^6Li to Z = 25) /169/. In the region of ^{208}Pb, the pion is captured mainly from high orbits, e.g., 4f. While in 1s atomic capture, giant dipole levels are predominantly excited, it was noted that 2p capture leads to strong excitations of giant quadrupole resonances also /170/. This was confirmed, e.g., in ^{16}O by both Goldhaber-Teller model /165/ and shell-model calculations /121, 171/. In ^{208}Pb, there are strong captures leading to even higher angular momentum states, such as the unnatural parity states $J = 4^-, 5^+, 6^-$, and 7^+. The results of shell-model calculations /162/ indicating such effects are displayed in Figure 24, together with averaged experimental results (dashed curve) /163/. It should be stressed that for p or higher wave pion capture the amplitude <u>cannot</u> be approximated by (104a), but the full amplitude has to be used (see next subsection).

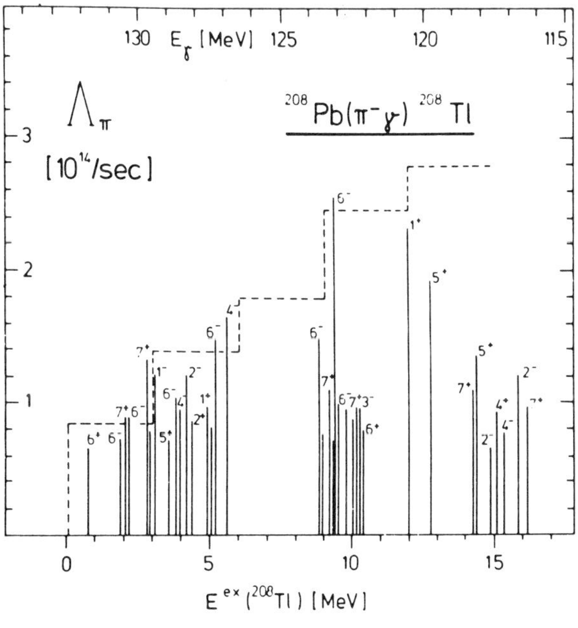

Fig. 24. Calculated excitation spectra of ^{208}Tl after radiative pion capture on ^{208}Pb /162/. The photon energy is given by the top scale. The dashed top line shows the measured photon spectrum /163/ averaged over 3 MeV intervals in units 10^{15}/s MeV

No measurement on radiative pion capture in flight (105b) have been carried out as yet, but calculations of the capture rates and angular distributions in ^{12}C have been performed using the Helm model /154/. Due to the $1/v_\pi$ kinematical factor of capture processes in flight, these experiments are possible due to the enhanced cross section close to threshold.

5.3 Pion Photoproduction

We shall here discuss nuclear pion photoproduction (103), restricting ourselves to the photoproduction of charged pions. No recent experimental π^0 photoproduction data are available, while some work along these lines is now in progress at the Bates-MIT linac /172, 173/. Although a considerable number of recent $\pi^\pm$ photoproduction experiments were carried out /174, 175/, they are mainly concerned with the excitation of low-lying levels of the daughter nucleus, below the giant resonance. Only a few of these experiments include the excitation of the giant resonance.

Restricting oneself to photon energies near the pion threshold, $E_\gamma^{thr} \cong m_\pi + \omega$, one is permitted by first approximation to use for the interaction Hamiltonian the KROLL-RUDERMAN term only /152/,

$$H_{int} = -i \frac{2\pi\sqrt{2}}{(E_\pi k)^{1/2}} \frac{ef}{m_\pi} \sum_{i=1}^{A} \tau_i^\pm \underline{\sigma}_i \cdot \underline{\varepsilon}\ \delta(\underline{r}-\underline{r}_i). \tag{106}$$

Here, $e^2 = 1/137$, $f^2 = 0.08$; $\underline{k}$ and E_π, m_π are the photon momentum and pion total energy and rest mass, respectively, and $\underline{\varepsilon}$ is the photon polarization vector. Additional interaction terms are momentum dependent and hence small near threshold /41/, and all except one of those also contain the spin operator $\underline{\sigma}_i$. It has been demonstrated, however, employing a Fermi gas model /176/, that the use of (106) is adequate even up to ~40 MeV of energy above threshold (with typical errors below ~20%), due to cancellations between the momentum-dependent terms.

With H_{int} of (106), and the use of plane wave pions, an early calculation of photopion production with excitation of the giant resonance in ^{16}O has been carried out by KELLY et al. /177/, employing the Goldhaber-Teller model. More recent work for ^{12}C /154/ and ^{16}O targets /144/ was based on the Helm model.

Improving on the simple interaction Hamiltonian of (106), calculations have been carried out using the full transition matrix of the elementary interaction due to BERENDS et al. /178/, plane /179/ and, more recently, distorted /144, 180-182/ pion wave functions.

In presenting some of these results, we shall first illustrate the qualitative aspects of the reaction of (103) using plane waves and the Kroll-Ruderman Hamiltonian, describing the nuclear form factors by the Helm model as discussed in Sect. 4.

Subsequently, the effects of the improvements due to the use of the full interaction an pion wave distortion will be considered, as well as the effects of the relativistic transformation from the center-of-mass system of the elementary process to the laboratory system, and also some corrections on the distorted wave approximation.

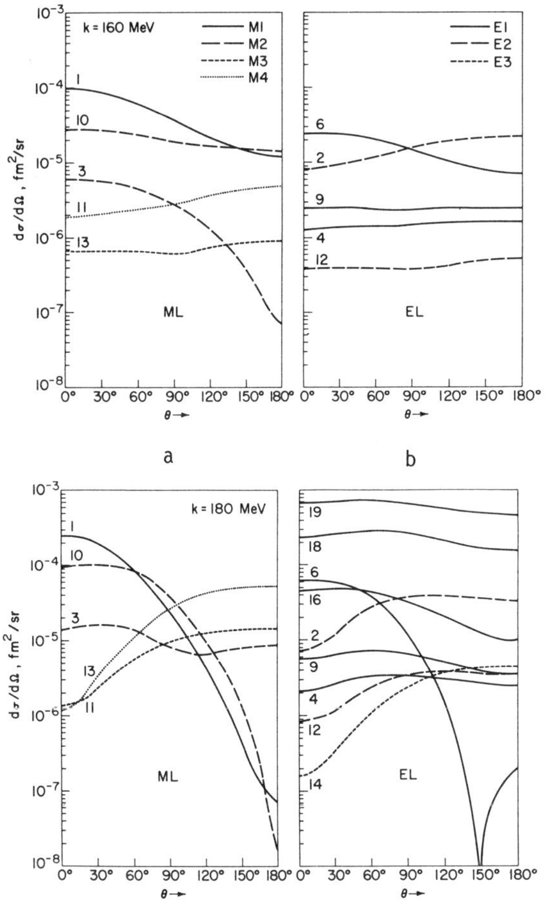

Fig. 25. Differential cross section of π^+ photoproduction on a ^{12}C target, with excitation of a) magnetic and b) electric states plotted vs pion emission angle θ /154/

For a ^{12}C target, Figure 25a,b present the angular distributions of photoproduced π^+ mesons at photon energies k = 160 and 180 MeV, with the excitation of magnetic (a) and electric (b) analog levels as shown in Figure 17, using the Helm model with parameters of Table 5 (except for levels No. 1-4 for which the parameters of /66/ were taken). Plane wave pions and the interaction of (106) were used here. The giant resonances correspond to the curve numbered 18 or 19, and the giant M1 resonance is state No.1.

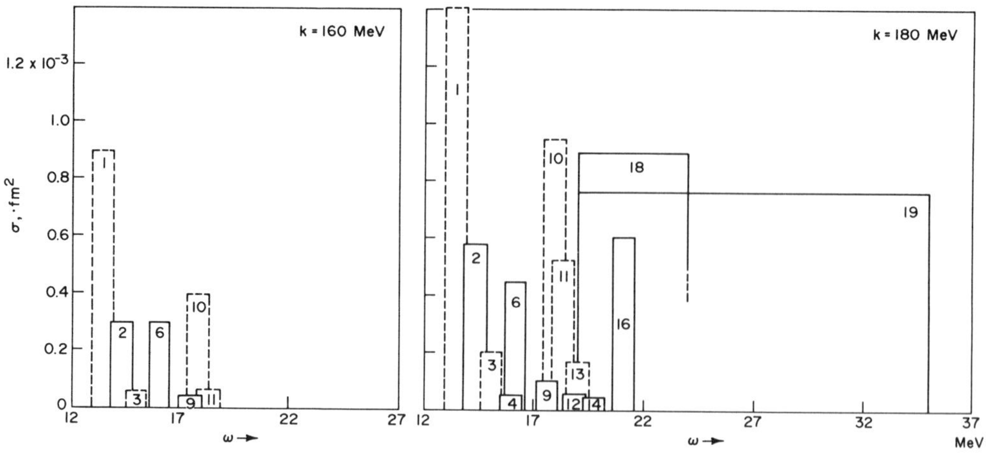

Fig. 26. Total π^+ photoproduction cross sections on a ^{12}C target, plotted vs excitation energy ω. Left portion: incident photon energy k = 160 MeV; right portion: k = 180 MeV. States are numbered according to Table 5 /154/

One should note that if the incident photon energy becomes very low, it is possible that some high-lying states (e.g., giant resonances) may no longer get excited because of the lack of available energy, or they may become strongly suppressed. This is also illustrated in Figure 26, where we plotted the total π^+ photoproduction cross section (integrated over all pion emission angles) at k = 160 MeV (left portion) and k = 180 MeV (right portion), vs the excitation energy ω, assigning each state an arbitrary width of 1 MeV except the giant resonances (nos. 18 and 19) which were given the widths shown in Table 5. It is seen that at 160 MeV, only the low-lying excited states show up, so that the photoproduction process may provide a means for studying the excitation of bound states without the overwhelming beckground of the giant resonance. It is possible, of course, that the tails of the giant resonances, albeit in a suppressed way, may still appear even at lower energies; this effect has not been included in the figure and deserves further scrutiny. Similar results were obtained for ^{16}O /144/ and are shown in Figure 27, with levels numbered as in Table 6.

Photopion production contributes to total photoabsorption cross sections in nuclei around the pion threshold. These measured cross sections /118/ indeed show a rise above ~140 MeV, as illustrated in Figure 28 for a ^9Be target. The theoretical curves in the figure were obtained in a calculation /154/ of giant resonance excitation during π^+ photoproduction using the Goldhaber-Teller model, together with coherent π^0 photoproduction using the nonspin-flip term of the BERENDS interaction

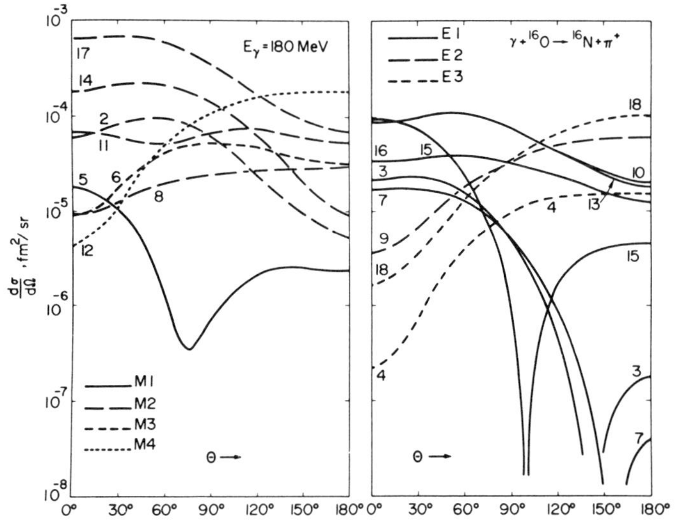

Fig. 27. Differential cross section of π^+ photoproduction on a ^{16}O target, with excitation of magnetic (left portion) and electric states (right portion) plotted vs pion emission angle θ /144/. Plane wave pions and Kroll-Ruderman approximation were assumed here

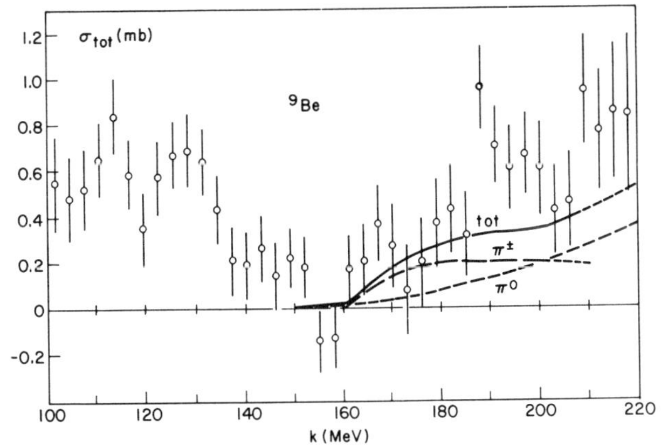

Fig. 28. Total photoabsorption cross section in ^{9}Be at the pion threshold /154/. Data are from /118/

Hamiltonian /178/,

$$2\pi \sum_{i=1}^{A} \left(1 + \frac{m_\pi}{m}\right) D^{(o)} \underline{\varepsilon} \cdot (\underline{p}_\pi \times \underline{k})\, \delta(\underline{r} - \underline{r}_i). \qquad (107)$$

It is seen that most of the rise in the cross section may be accounted for by the giant resonance excitation mechanism of pion photoproduction.

A plane-wave pion calculation which used the full interaction amplitude together with plane-wave pions treated the topic of isospin effects in pion photoproduction /179/ (similar to the discussion in Sect. 1.4) for photoabsorption and was based on the same 2 particle-1 hole shell model /64/.

The interesting feature is here that while π^- photoproduction leads to both the T = 1/2 and T = 3/2 levels in ^{13}N, π^+ photoproduction leads to the levels of

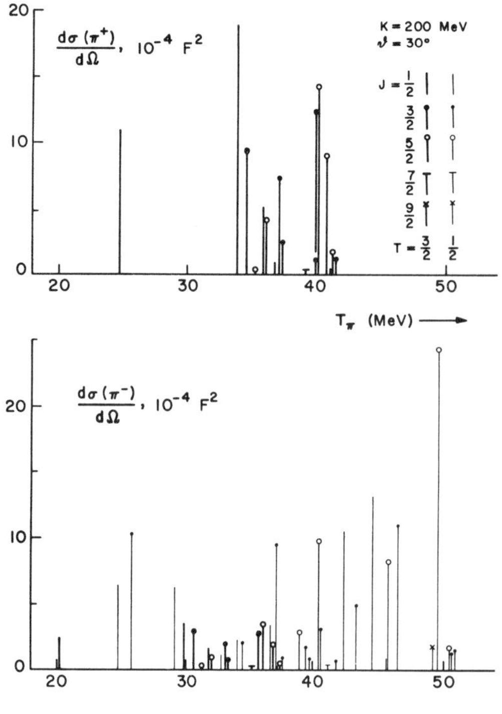

Fig. 29. Spectra of π^+ and π^- mesons photoproduced from ^{13}C with excitation of T = 1/2 (light lines) and T = 3/2 (heavy lines) analog levels in ^{13}B and ^{13}N, respectively (photon energy k = 200 MeV) /179/

^{13}B which are exclusively T = 3/2. In this way, i.e., by comparing the π^+ and π^- photoproduction spectrum (see Figure 29), one may obtain direct, unambiguous information on the isospin splitting of the giant resonance, while this information is less unambiguous if electron scattering is used.

One may at this point give consideration to the effects of the final-state interaction between pion and nucleus which distorts the pion wave function. The expression for the cross section of the reaction Eq. (103) is

$$\left(\frac{d\sigma}{d\Omega}\right)_{\gamma\pi} = \frac{p_\pi}{km_\pi^2}\,\frac{e^2f^2}{\hat{J}_o^2} \sum_{M_oM_fs_\gamma} |\underline{\varepsilon} \cdot \mathcal{M}|^2 \tag{108}$$

containing sums over magnetic quantum numbers of initial (J_o, M_o) and final (J_f, M_f) nuclear states, and over photon polarization s_γ. Kinematics gives for the momentum transfer

$$\underline{q} = \underline{p}_\pi - \underline{k} \tag{109a}$$

and for the pion energy

$$E_\pi = k - \omega - (q^2/2Am), \tag{109b}$$

where $\underline{p}_\pi$ and $\underline{q}$ are the pion and nuclear recoil, respectively, and ω is the excitation energy of the final nuclear level measured from the ground-state of the initial nucleus.

The nuclear matrix element in the Kroll-Ruderman limit contains the spin-flip transition operator (106), and for plane-wave pions is given by /154/

$$\underline{\mathcal{M}}\,(\underline{q}) = \langle J_fM_f|\sum_{j=1}^{A} \underline{\sigma}_j\tau_j^{\pm}\, e^{-i\underline{q}\cdot\underline{r}_j}|J_oM_o\rangle. \tag{110a}$$

The photon polarization sum becomes simply

$$\sum_{s\gamma} |\underline{\varepsilon} \cdot \underline{\mathcal{M}}| = |\mathcal{M}|^2 - |\hat{k} \cdot \mathcal{M}|^2, \tag{110b}$$

where $\hat{k} = \underline{k}/k$. The matrix element may be expressed by a Fourier transform of the transition density,

$$\underline{\mathcal{M}}\,(\underline{q}) = \int d^3r\, e^{i(\underline{k}-\underline{p}_\pi)\cdot\underline{r}}\, \underline{\rho}(\underline{r}), \tag{110c}$$

where the density

$$\underline{\rho}(\underline{r}) = \langle J_fM_f|\,\underline{\rho}^{op}(\underline{r})\,|J_oM_o\rangle \tag{110d}$$

as defined in (77) is given in terms of the transition operator $\underline{\rho}^{op}(\underline{r})$ of (78), and contains the isospin operator $\tau_i^{\pm}$.

The final-state interactions of the pion with the nucleus can be included in (110a) by replacing the wave function $\exp(-i\underline{p}_\pi \cdot \underline{r})$ of the plane-wave pion by that of a distorted-wave pion, obtained as a solution of the Klein-Gordon equation

$$(\underline{p}_\pi^2 + m_\pi^2)\ \phi = \left[(E_{tot} - V_{coul} - Am - \underline{p}_\pi^2/2Am)^2 - 2E_\pi V_{opt}(r)\right] \phi \tag{111}$$

containing a pion-nucleus optical potential. In phenomenological calculations the latter often is assumed to have the structure for $\pi^{\mp}$ mesons /183/

$$\begin{aligned} V_{opt} = &- (2\pi/E_\pi) \Big\{ (1 + \frac{E_\pi}{m}) \left[b_o \rho(r) \pm b_1(\rho_n - \rho_p) \right] \\ &+ \frac{E_\pi}{2m} \nabla^2 C'(r) + B(1 + \frac{E_\pi}{2m})\ \rho^2(r) \\ &- (1 + \frac{E_\pi}{m})^{-1}\ \underline{\nabla} \cdot \frac{C'(r)}{1 + \frac{4\pi}{3} \xi C'(r)} \underline{\nabla} \\ &- C(1 + \frac{E_\pi}{2m})^{-1}\ \underline{\nabla} \cdot \rho^2(r)\ \underline{\nabla}, \end{aligned} \tag{112a}$$

where

$$C'(r) = c_o \rho(r) \pm c_1 \left[\rho_n(r) - \rho_p(r)\right], \tag{112b}$$

$\rho(r)$ being the sum of the nuclear proton $[\rho_p(r)]$ and neutron $[\rho_n(r)]$ static densities. The values of the complex parameters b_o, b_1, c_o and c_1 are energy dependent. They can be obtained from measured pion-nucleon scattering phase shifts or from pionic-atom data /184/; the latter will also provide the coefficients B, C of the pair interaction terms in V_{opt}. The term containing the coefficient $0 \leqq \xi \leqq 1$ is referred to as the Lorentz-Lorenz effect /185, 186/ of pions in nuclear matter. The pion wave function itself is expanded in the form

$$\begin{aligned} \phi^{(-)*}(\underline{p}_\pi, \underline{r}) &= \phi^{(+)}(-\underline{p}_\pi, \underline{r}) \\ &= 4\pi \sum_{l\,m} i^l e^{i(\delta_l + \sigma_l - \sigma_o)} R_l(r)\ Y_{lm}(\hat{r})\ Y^*_{lm}(-\hat{p}_\pi), \end{aligned} \tag{113}$$

where σ_l is the Coulomb phase shift, and δ_l is the pion-nucleus strong interaction phase shift.

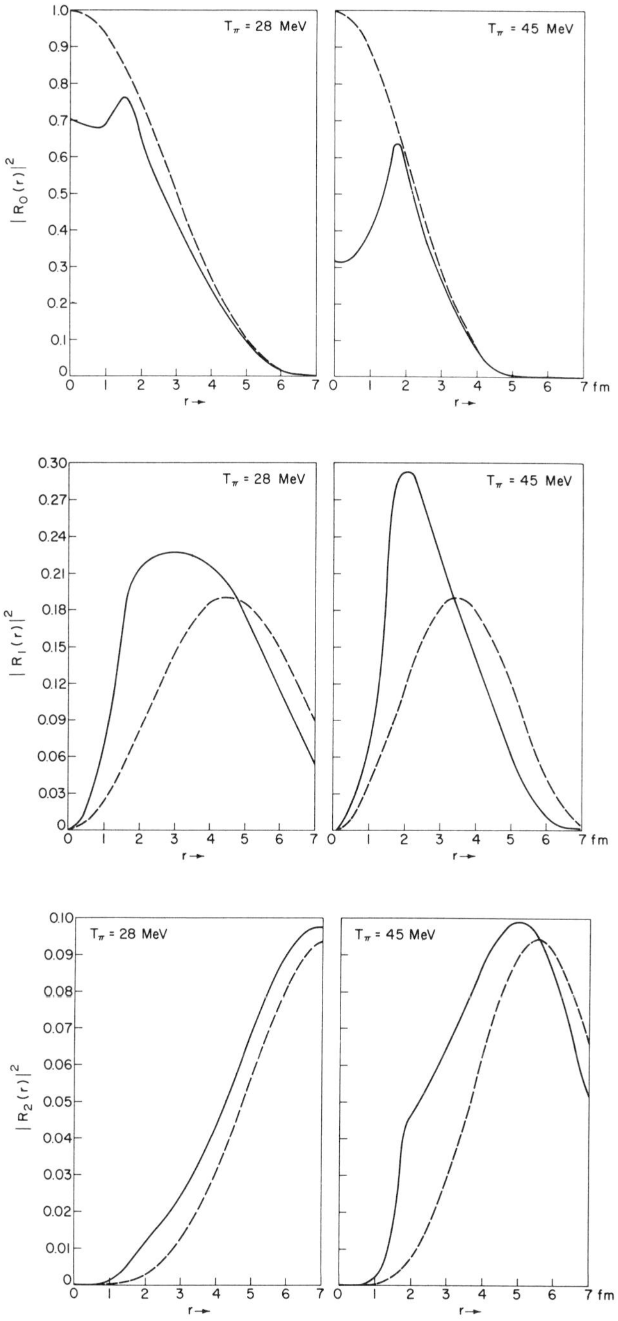

Fig. 30. Absolute square of radial π^- wave function for distorted (solid curve) and free case (dashed curve) at pion kinetic energies T_π as indicated, for l = 0 (top), 1 (center) and 2 (bottom) /154/ in ^{12}C

In Figure 30 we plot the absolute squares of the radial π^- wave functions $R_l(r)$, $l = 0, 1, 2$ (solid curves) together with the squares of the corresponding spherical Bessel functions $j_l(p_\pi r)$ (dashed lines), for pion kinetic energies T = 28 MeV and 45 MeV as obtained in /154/. The s wave is seen to be suppressed near the origin due to absorption and to the repulsive s wave part of the optical potential which overwhelms the attractive effect of the Coulomb potential. The p and d waves are enhanced due to the attractiveness of the optical potential. Since the transition mainly takes place at the nuclear surface (e.g., the Helm model), i.e., at a radius R ~ 2-3 fm for ^{12}C, the corrections due to pion distortions are seen from Figure 30 to amount to ~20% for the π^- nucleus s wave contribution. The higher waves however, may have corrections of the order of 100%.

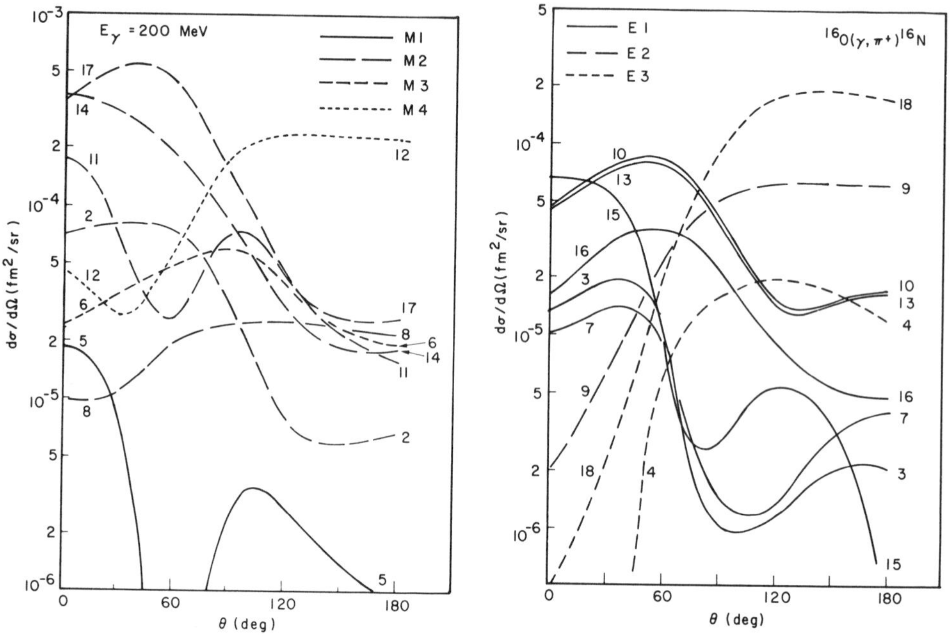

Fig. 31. Differential cross section of π^+ photoproduction on ^{16}O, with excitation of magnetic (left portion) and electric states (right portion) plotted vs pion emission angle θ. Distorted waves and full Berends amplitudes are used in the calculation /143/

The angular distributions of positive photopions from ^{16}O, calculated with distorted pion wave functions (and the full interaction) are shown in Figure 31 /143/, to be compared with Figure 27 obtained with plane-wave pions and the Kroll-Ruderman term.

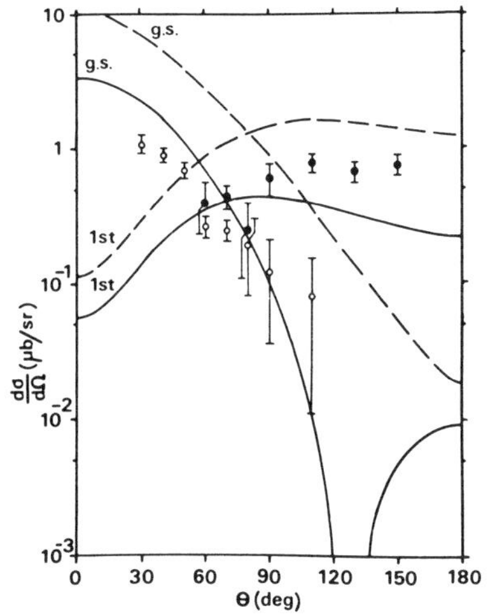

Fig. 32. Pion angular distributions in ^{12}C (γ, π^+) ^{12}B at k ≅ 194 MeV leading to the ground-state (open circles) and the first excited state (full circles) in ^{12}B /187/. Solid and dashed curves are theoretical Helm and j-j shell-model plane-wave estimates with the Kroll-Ruderman term

Quantitative measurements have so far been carried out for excitations to the low-lying levels of light nuclei only. Figure 32 presents data on ^{12}C (γ, π^+), obtained by unfolding a virtual photon spectrum from the measured ^{12}C $(e, e' \pi^+)$

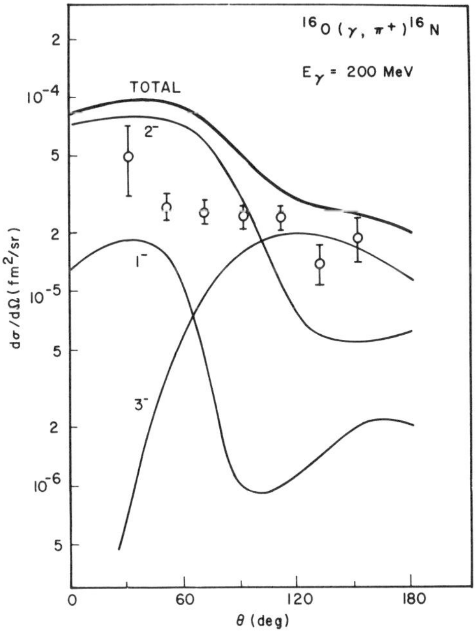

Fig. 33. Pion angular distributions in ^{16}O (γ, π^+) ^{16}N at k = 200 MeV leading to the lowest states of ^{16}N. Data from /191/, theory from /143/, i.e., distorted-wave impulse approximation using full Berends amplitude and the Helm model for ^{16}N

process /187/, showing pion angular distributions to the ^{12}B ground and first excited state. As expected, the plane-wave calculations agree much better with the Helm model results /154/ (solid curves) than with the results of a pure j-j shell-model calculation /188/ (dashed curves), since the latter is known to generally overestimate the spin-flip contribution to the ^{12}B ground-state by factors of about 5 /189/ while the former, being phenomenologically fitted to (e, e') form factors, reproduces the experimental results. Figure 33 presents (γ, π^+) angular distributions from ^{16}O summed over the four lowest levels of ^{16}N /143/, compared to theoretical angular distributions (for the 1^-, 2^-, and 3^- levels only, since that of the 0^- level is negligible), using distorted pion waves. In Figure 34, the integrated theoretical cross sections as functions of k are compared with data of /190, 191/. The full Berends interaction has been used in these calculations /143/.

Excitation of the giant resonances in the $(\gamma, \pi^\pm)$ reactions have so far been measured only at a few occasions. In the "isochromate"curves of Figure 35,

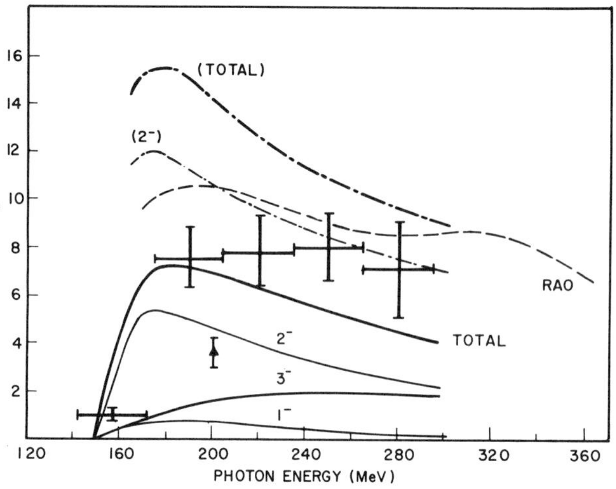

Fig. 34. Total photopion cross sections in ^{16}O (γ, π^+) ^{16}N as a function of photon energy k, leading to the lowest states of ^{16}N. Data from (190, 191/, theory from /143, 192/. Solid curves: distorted wave impulse approximation using the full Berends amplitude and the Helm model for levels 2, 3, 4, of Table 6 and their total. Dashed-dot curve: plane-wave calculation. Dashed curve: shell-model calculation from /192/

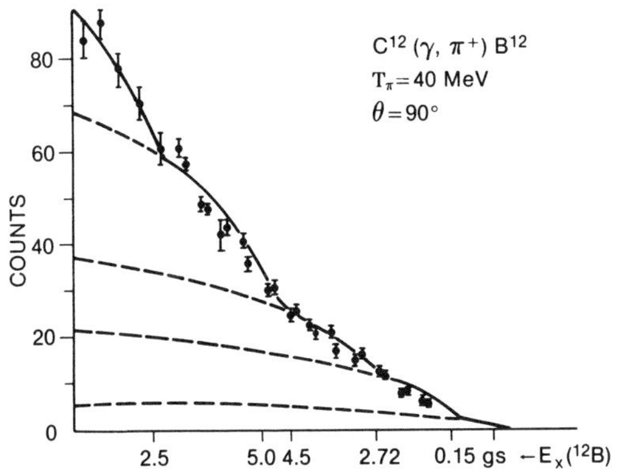

Fig. 35. MIT-Bates "isochromate" curves of the ^{12}C (γ, π^+) ^{12}B reactions, obtained at a pion emission angle of 90° with pion energy T_π = 40 MeV /159/, plotted vs the ^{12}B excitation energy

obtained at the MIT-Bates linac /159/ where the emitted π^+ was observed at a constant energy T_π = 40 MeV while the incident photon (bremsstrahlung endpoint) energy was varied, the contributions of the ^{12}B excited states are seen to come in consecutively, with the giant E1 resonance entering at ≧7 MeV excitation in ^{12}B.

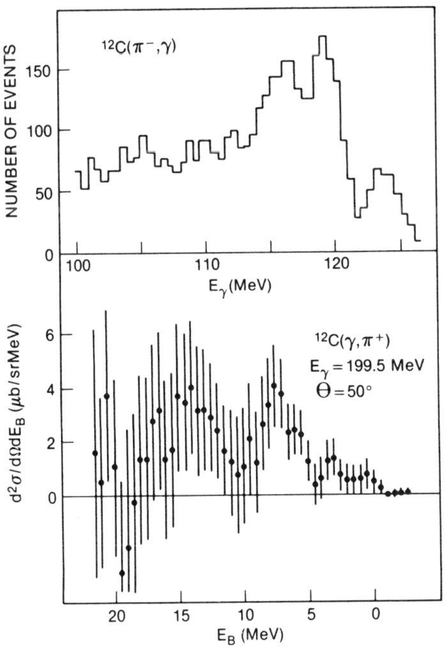

Fig. 36. Excitation cross section of ^{12}B in ^{12}C (γ, π^+) ^{12}B (bottom) /191/, compared to excitation in the inverse reaction ^{12}C (π^-, γ) ^{12}B (top) /193/, plotted vs ^{12}B excitation energy

SUNG et al. /191/ obtained the excitation cross section of ^{12}B in the same reaction at 50° as shown in the lower portion of Figure 36; this is compared to the excitation in the inverse reaction $^{12}C\ (\pi^-, \gamma)\ ^{12}B$ (upper portion). The spin-flip E1 strength seems here to be concentrated between 10 and 15 MeV in ^{12}B, which corresponds to 23-28 MeV in ^{12}C.

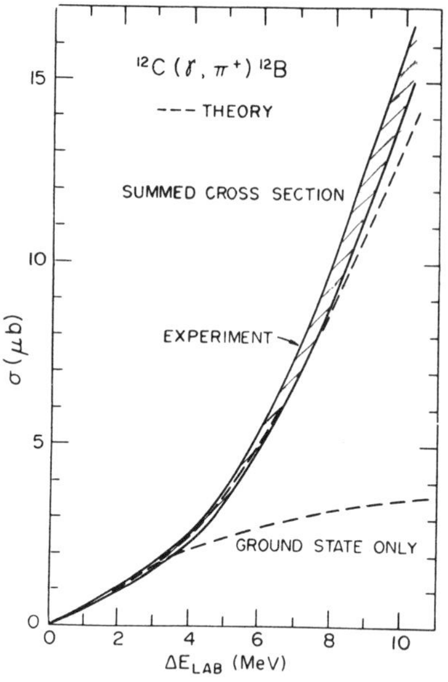

Fig. 37. Cross section summed over states in ^{12}B from $^{12}C\ (\gamma, \pi^+)\ ^{12}B$, integrated over pion angles and plotted vs the maximum excitation energy ΔE, with theoretical data points /182/

Another measurement of the ^{12}B excitation spectrum in $^{12}C\ (\gamma, \pi^+)\ ^{12}B$, summed over the contributions of all levels up to the maximum excitation energy ΔE and integrated over pion angles (by detecting the reaction via an observation of the $\pi^+ \to \mu^+, e^+$ decay chain) was carried out by MILDER et al. /182/. Figure 37 shows the data (contained within the band between the solid curves), with theoretical points obtained by a calculation using the full Berends interaction and pion waves distorted by the optical potential of (112). Again, the giant resonance appears above ≧8 MeV.

We shall conclude this section with a few remarks referring to open problems in the distorted-wave impulse approximation (DWIA). As in π-nucleus scattering, a general problem is presented by the required transformation of the π-nucleon center of mass amplitude to the π-nucleus center of mass amplitude (in which the distorted pion wave is calculated). A large literature exists on this subject (see, e.g., /184/). Conceptually, the simplest procedure is given by going back to the original covariant form /178/ for the amplitude and reducing it according to (70). Employing it in different Lorenz systems /144/, one can impose the condition $\underline{\varepsilon}\cdot\underline{k}=0$ in both systems and obtain the transformation for the coefficients $\mathcal{F}_i$ which enter in the full amplitude

$$\mathscr{F} \equiv i\,\underline{\sigma}\cdot\underline{\varepsilon}\,\mathscr{F}_1 + \underline{\sigma}\cdot\underline{p}_\pi\,\underline{\sigma}\cdot(\underline{k}\times\underline{\varepsilon})\,\mathscr{F}_2/p_\pi k$$
$$+ i\,\underline{\sigma}\cdot\underline{k}\;\underline{p}_\pi\cdot\underline{\varepsilon}\,(\mathscr{F}_3/p_\pi k) + i\,\underline{\sigma}\cdot\underline{p}_\pi\;\underline{p}_\pi\cdot\underline{\varepsilon}\,(\mathscr{F}_4/p_\pi^2). \tag{114}$$

The mentioned Lorenz transformation leads to the same amplitude in both π-nucleon and π-nucleus C.M. systems, but with different coefficients $\mathscr{F}_i$, if the initial nucleon momenta have here been set = 0 or $-(\underline{p}_\pi-\underline{k})/2$, which are two possible choices (the latter preferable) for approximating the nuclear Fermi motion. Obviously, the kinematics of the participant particles in the elementary process is different depending on whether pion photoproduction takes place on a free nucleon or on a nucleon bound in a nucleus. This causes the thresholds of the two reactions to differ depending on the pion emission angle and on the nuclear excitation energy (109b). When utilizing the tabulated amplitudes $\mathscr{F}_i$ of the elementary process /178/, one may, e.g., arrive at their threshold value for a pion energy above the actual threshold of the reaction on a nuclear target. To illustrate the ambiguities arising in this context /144/, we mention the fact that for a fixed pion emission angle, there might in general (besides the pathological case of subthreshold photons) exist two different on-shell processes such that either the pion or the photon energy is equal to the corresponding energy in the nuclear reaction [cf. (109b)].

Besides these kinematical effects, there are possible dynamical effects which have to be analyzed more carefully. For the distorted-wave impulse approximation, one neglects the role of possible (π, π') reactions in the final-state interaction, but only considers elastic rescattering on the state excited by the (γ, π) reaction. Furthermore, there are difficulties with DWIA in any reaction with a complex nucleus where the specific projectile-target nucleon interaction that produces the transition under consideration also contributes to the distortion of the projectile wave /194/. In our case, the $\mathscr{F}_i$ coefficients contain the πN phase shifts corresponding to the final-state interaction of the pion with the nucleon on which it has been produced. On top of all these problems, there is still the question of how the elementary operators themselves get modified in a nuclear environment (binding corrections, etc.).

6. Higher-Multipole Giant Resonance Excitation by Hadrons

6.1 Hadron Probes and Nuclear Structure

As shown in the previous sections, one may discern among the numerous processes occurring in the interaction of elementary particles with nuclei at intermediate energy a certain type of reactions (such as photoreactions up to incident energies 25-30 MeV, electroexcitation at low momentum transfer, muon capture, radiative pion capture, etc.) which excite the nucleus in a coherent fashion. This feature described, e.g., by the generalized Goldhaber-Teller model, is the excitation of giant resonances by the action of an external probe. The excitation operators can be classified for any orbital multipolarity as (see Sect. 1.4):

1) Isospin mode
2) Spin-isospin mode
3) Spin wave
4) Compressional mode (or density mode).

In the shell model, this can be rephrased saying that the external probe induces the formation of particle-hole (p-h) states. The residual interaction between the particle and the hole under these conditions results in the formation of coherent nuclear states which appear as giant resonances /62/. The structure of the external field acting on a nucleus manifests itself in the specific type of collective nuclear motion generated, characterized by angular momentum J, parity Π, and isospin T.

As we have previously discussed (Sect. 1), the appearance of the giant resonance in the photo- and electroexcitation of nuclei is a well-known example of this process.

More recently, convincing experimental evidence for the excitation of the giant resonance states in muon capture /34/ and in radiative pion capture /160/ has been obtained.

In the case of the "new" or "higher multipole" (going beyond the "old" electric or magnetic dipole), i.e., monopole, quadrupole, octupole, etc., resonances considered in Sect. 1.4, nearly all of the experimental evidence has been provided mainly by inelastic scattering of medium energy electrons and nuclear projectiles /8/ rather than by photonuclear reactions traditionally used to study the conventional giant resonance. Prior to this, the existence of the higher-multipole resonances was inferred from the requirement for effective charges in shell-model calculations of electromagnetic transition rates between low lying states caused by virtual giant resonance excitation, and from the core polarization renormalization in the microscopic description of inelastic nuclear scattering /62, 195, 196/.

As we discussed in Chapter 2 it has been found that high spin states tend to dominate the direct nuclear response at high momentum transfer. This phenomenon is attributable to the basic properties of Bessel functions of higher order and angular momentum selection rules.

Hadron scattering in general involves a comparatively high momentum transfer. This fact presents one advantage of studies of giant resonance excitation in hadron scattering experiments since, given the ability to unfold the data, information about higher multipoles can be obtained and compared with results obtained in electron scattering. Furthermore, with the variety of projectiles available for nuclear scattering, different types of giant resonances can be selectively excited. Specifically in proton, ^{3}He or ^{3}H scattering, all possible electric and magnetic isoscalar and isovector excitations are allowed. Deuteron scattering preferably excites electric and magnetic isoscalar resonances and α scattering essentially excites only isoscalar electric resonances. The pion probe which we will deal with at the end of this section may, through double charge exchange, $\pi^{\pm} \rightarrow \pi^{\mp}$, also generate isotensor excitations.

6.2 (p,p') Reactions

We shall first examine medium energy nucleon-nucleus scattering. To illustrate the kind of information which may be obtained by considering transitions to particular excited levels, we shall initially consider a simplified model of the interaction. Subsequently, we shall discuss some of the refinements which must properly be taken into account in a correct interpretation of (p,p') scattering.

We shall describe the use of plane-wave impulse approximation (PWIA) and of a zero-range, spin and isospin dependent interaction; furthermore, we shall ignore antisymmetrization effects between the assumed closed-shell ($J_0 = T_0 = 0$) target and the projectile.

The zero-range interaction responsible for inelastically scattering the nucleon and exciting the target to a p-h state is assumed to be /197/

$$V = \sum_{j=1}^{A} \left\{ A + B\underline{\sigma}\cdot\underline{\sigma}_j + C\underline{\tau}\cdot\underline{\tau}_j + D\underline{\sigma}\cdot\underline{\sigma}_j\ \underline{\tau}\cdot\underline{\tau}_j \right\} \delta(\underline{r}-\underline{r}_j) \qquad (115a)$$

(with $\underline{\sigma}$, $\underline{\tau}$ referring to the projectile nucleon).

In PWIA, this potential leads to a transition matrix element

$$t_{of} = \langle J_f | \sum_{j=1}^{A} (A + B\underline{\sigma}\cdot\underline{\sigma}_j + C\underline{\tau}\cdot\underline{\tau}_j + D\underline{\sigma}\cdot\underline{\sigma}_j\underline{\tau}\cdot\underline{\tau}_j) e^{i\underline{q}\cdot\underline{r}_j} | J_o \rangle, \tag{115b}$$

where the factor $\exp(i\underline{q}\cdot\underline{r}_j)$ admits a straightforward multipole expansion, which may couple with the $\underline{\sigma}_j$ operators leading to standard operators similar to $\mathcal{M}$ (q) or $\underline{\mathcal{M}}$ (q) as defined in Chapters 4 and 5.

The states excited by the spin-isospin independent potential $\sum_j A\,\delta(r-r_j)$ have $\Pi = (-1)^J$, T = 0. Analogously, $\sum_j C\underline{\tau}\cdot\underline{\tau}_j\,\delta(\underline{r}-\underline{r}_j)$ excites states $\Pi = (-1)^J$, T = 1 in reactions such as (p,p') (n,p) (p,n) (n,n'), etc. The high spin levels, e.g., 6^-(^{24}Mg) and 8^- (^{58}Ni), non-normal parity states which were discussed in Chapter 2 as ML(T=1) in connection with high momentum transfer electron scattering can (in this simple model) only be reached via the spin-isospin dependent potential

$$\sum_j D\underline{\tau}\cdot\underline{\tau}_j\underline{\sigma}\cdot\underline{\sigma}_j\,\delta(\underline{r}-\underline{r}_j).$$

The traditional approach to studying medium energy "direct" reactions of a strongly interacting probe with a nucleus is the distorted-wave impulse approximation (DWIA). The inelastic transition is viewed to occur as a quasi one-step process described by a finite range potential, with multiple scattering effects being confined to the distortion of incoming and outgoing projectile wave functions. As in the simple model we described before, the inealstic transition is thus induced by an effective interaction which is a sum of one-body operators in the space of the target nucleons, which may be considered as a pseudopotential /26/. Concerning the question whether the distortion of the probe, and use of a more realistic effective interaction, destroy the simple relationship of the mentioned features of (p,p') or (p,n) with those of electron scattering, MOFFA and WALKER /197/ found that the results of distorted-wave and plane-wave analysis do not differ appreciably for 183 MeV protons.

The crucial point is that distortion does not drastically change the shapes of the angular distribution nor the relative strengths of the differential cross sections of medium-energy nucleons ($E_p \gtrsim 100$ MeV) for the different excited states. Thus a prediction that the same high spin giant multipole states strongly excited at large momentum transfer in inelastic electron scattering will be seen in proton scattering, and in particular in charge exchange scattering (leading to T = 1 levels) can be made, see Figure 38. This prediction also depends on the relative strengths of the different terms in the effective interaction. If these spin and isospin dependent interactions were relatively weak, then the excitation of these states could be weak even though the comparison with the electron scattering formalism would still be valid. The spin and isospin dependent terms are quite

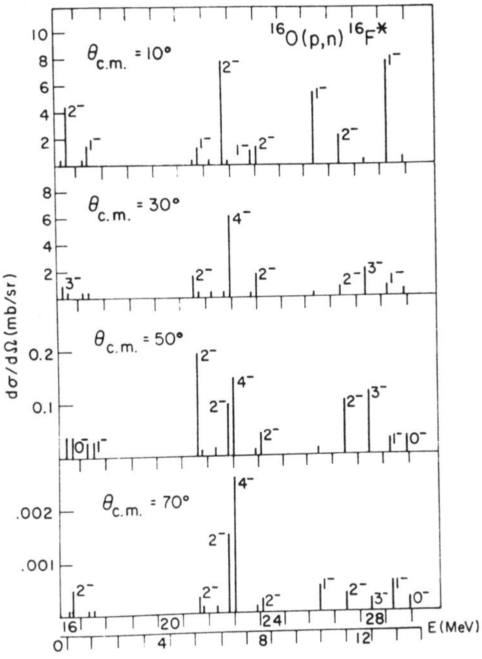

Fig. 38. ^{16}O (p,n) cross section for several scattering angles at E_p = 183 MeV. The upper energy scale gives the excitation energy of these states relative to the ^{16}O ground-state and the lower scale relative to the ^{16}F ground-state /197/

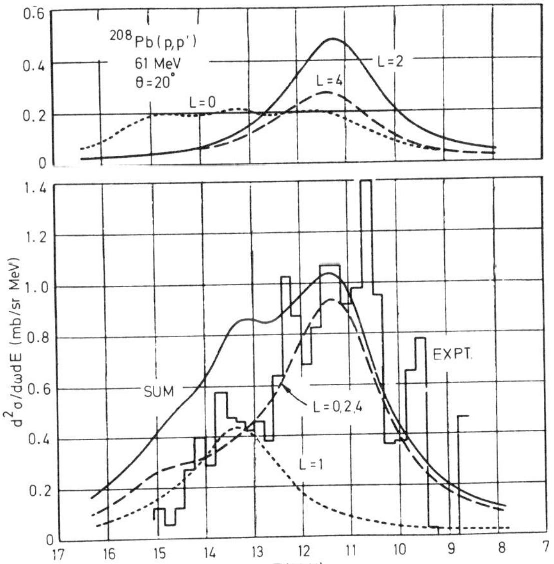

Fig. 39. The giant resonance peak in the spectrum of E_p = 61 MeV protons scattered from ^{208}Pb at a (c.m.) angle of 20°. The various curves show theoretical (shell-model) predictions /202/ for the contributions from the different multipoles L. The experimental cross section results after subtraction of a uniform background of 3.7 mb/sr MeV from the measured spectrum

strong, however; this is a manifestation of the fact that the basic nucleon-nucleon interaction itself has large spin and isospin dependent components. On the other hand (neglecting antisymmetrization), T = 0 states are not strongly excited since the isospin independent part is weak /198/.

This conclusion can change, however, once exchange terms are taken into account /198/. One such state, a 6^-, T = 0 state in ^{28}Si, has recently been seen experimentally /199/; performing both (p,n) and (p,p') experiments, certain states appearing in both spectra could be positively identified as T = 1 states.

At lower energies (≲50 MeV), the simple description based on a one-step process alone does not suffice any longer /200/. One has also to consider processes in which the first step involves the capture of the projectile; the intermediate excited state reached in this step then decays (semidirect amplitude) and the residual nucleus is left in the definite (excited) final state. This amplitude is strongly dependent on the energy of the projectile and has all the features of a resonance. The process may be of special importance at backward angles where it leads to prominent peaks. An analysis following these lines has recently indicated evidence of octupole giant resonance strength in the 30-40 MeV excitation region at E_p = 42.5, 44.0, 49.25 MeV in p,p' reactions on ^{16}O /201/.

Although the experimental techniques used to study direct high-excitation resonances are generally the same as those used for low-lying levels, inelastic scattering studies at high excitation energies do present some special problems. The resonance cross section is only a small fraction of the total inelastic cross section in the resonance region and assumptions about the nature of the spectrum underlying the resonance peak must be made (the shape and magnitude of the underlying continuum must be obtained by extrapolation, see Figure 39 /202/).This figure shows one of the first examples of the "new giant resonances", including the T = 0, L = 2 giant quadrupole resonance (GQR) which is found at ~2 MeV below the conventional T = 1, L = 1 giant dipole resonance, as described in recent reviews /8/.

6.3 (α, α') Reactions

Reactions of inelastic alpha particle scattering may be useful to extract isoscalar strengths, to be compared with isoscalar electromagnetic strengths /203/. In this context, the sum rules most often used for such a comparison are the energy weighted ones for the operators $r^L Y_L^M(\theta, \phi)$. Consider the isoscalar (T = 0) and the isovector (T = 1) operators Q_{TLM}

$$Q_{0LM} = \sum_i r_i^L Y_L^M(\theta_i, \phi_i) \tag{116a}$$

$$Q_{1LM} = \sum_i \tau_{3i} r_i^L Y_L^M(\theta_i, \phi_i). \tag{116b}$$

In a similar fashion, neutron and proton operators can be defined

$$Q_{pLM} = \frac{1}{2} (Q_{0LM} + Q_{1LM}) \tag{117a}$$

$$Q_{nLM} = \frac{1}{2} (Q_{0LM} - Q_{1LM}). \tag{117b}$$

The proton operator is needed to describe EL electric transitions (note that we deal with $L > 1$ transitions where center-of-mass effects are not as large as in the dipole case, see Chapter 1). Assuming a spin zero ground state, the energy-weighted sum

$$\sum_L^T \equiv \sum_n \omega_n \sum_M |\langle n,L,M|Q_{TLM}|0,0,0\rangle|^2 \tag{118}$$

can be evaluated in an approximately model-independent way in the absence of velocity-dependent forces for $T = 0$ as well as of exchange forces for $T = 1$, see Sect. 2.3 and Appendix A,

$$\sum_L^0 = \frac{1}{8\pi m} A\, L(2L+1)^2 \langle r^{2L-2}\rangle_0. \tag{119a}$$

For a uniform distribution of radius R, one has

$$\langle r^{2L-2}\rangle = 3R^{2L-2}/(2L+1). \tag{119b}$$

The sum rule for EL transitions is similarly

$$\sum_L^P = \frac{1}{8\pi m} Z\, L(2L+1)^2 \langle r^{2L-2}\rangle_{\text{proton distribution}} \tag{120a}$$

but the radial average is now taken over the ground-state proton distribution. Only if this radial average is the same for neutrons and protons, then

$$\sum_L^P = \frac{Z}{A} \sum_L^0 \tag{120b}$$

and one can usefully compare (α, α') strengths with electric strengths. A more careful isospin analysis is needed in general especially for $N \neq Z$ nuclei.

As a typical example we mention the isoscalar quadrupole strength which has recently been investigated in several nuclei /8, 204/. For more detailed information about (α, α') processes and higher multipolarities (and the monopole breathing mode) we refer to the review by BERTRAND /8/. About the latter mode, we only recall that such a state indeed has been seen in forward-angle (<5°) α, α' scattering at E_α = 96 MeV on ^{144}Sm and ^{208}Pb /205/ (see Figure 40).

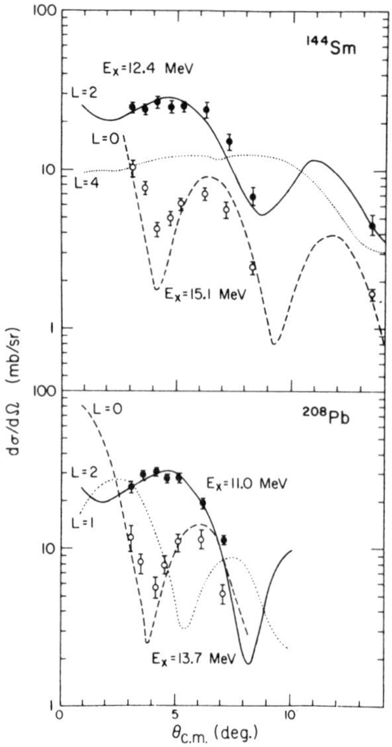

Fig. 40. Angular distributions of (α, α') reactions with excitation of T = 0 monopole and quadrupole components of the giant resonance in ^{144}Sm (12.4 and 15.1 MeV) and in ^{208}Pb (11.0 and 13.7 MeV). Results of DWIA calculations are shown for various values of the multipolarity L /205/

As a general remark, however, it should be stated that the extraction of electromagnetic sum rule strengths from inelastic hadron scattering is model dependent /202/. The reason is that the long wavelength limit [see (4)] does not apply here, and the operators describing the hadron-nucleus (in our case, α-nucleus) interaction depend on the energy of the projectile and on the scattering angle. For extracting isoscalar strengths from (α, α') scattering, use is made in general of the collective model within a distorted-wave framework. The uncertainties are in general smaller at small angle scattering (Figure 40).

6.4 (π, π') Reactions

The interaction of pions with nuclei is a vigorously debated field. Since a good understanding of this subject is still in a process of development /206/, we do not intend to present here an extensive review of the subject, but shall limit ourselves to selected remarks.

Again, as in (p,p') processes, one might expect high spin state excitation of $J_0 = 0$, $T_0 = 0$ nuclei to dominate at large angles /207/. A peculiar feature of pion scattering is the importance of excitations of T = 0 states. As compared to electron scattering, one recalls the reason for T = 1 high spin states dominating the nuclear response in inelastic electron scattering as due to the fact that in the energy range of electrons now available, high momentum transfer implies relatively large scattering angles where the transverse form factor F_T dominates over the longitudinal contribution F_C (see Sect. 2). Since the nucleon isovector magnetic moment $\mu_p - \mu_n$ is more than five times larger than the isoscalar moment $\mu_p + \mu_n$, T = 1 states dominate the contribution of the transverse multipoles to high spin (thus, spin-flip) transitions.

For pions, the relative importance of the T = 0 spin-flip (p,h) states can be traced back to the spin-isospin dependence of the basic pion-nucleon interaction operator. For example, if the P_{33} amplitude dominates, with its corresponding spin-isospin structure

$$<\pi_\alpha | P_{33} | \pi_\beta> \cong \left(\delta_{\alpha\beta} - \frac{1}{3}\tau_\alpha\tau_\beta\right)\left\{3(\underline{k}_i\cdot\underline{k}_f) - (\underline{\sigma}\cdot\underline{k}_i)(\underline{\sigma}\cdot\underline{k}_f)\right\} \tag{121}$$

(where $\underline{k}_i$, $\underline{k}_f$ are the pion initial and final momenta, respectively), the $\Delta T = 0$ part of the interaction will be twice as strong as its $\Delta T = 1$ counterpart. The problem of the weight of $\Delta T = 0$ and $\Delta T = 1$ excitations is interesting both theoretically and experimentally /208/.

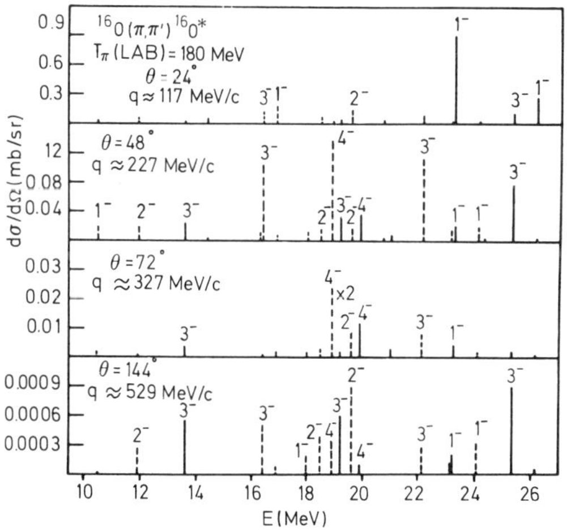

Fig. 41. Pion-^{16}O inelastic differential cross section as a function of final nuclear excitation energy. Solid lines correspond to T = 1 final nuclear states, while T = 0 states are represented by dotted lines, all calculated by using a particle-hole shell model /207/

Just as for (p,n) scattering, the pion charge exchange reaction on a T = 0 nucleus would not lead to any T = 0 excited state, and this allows experimentally a more positive identification of any excitation seen in inelastic pion scattering which may have been otherwise predicted to be T = 0 /207/.

Some results of p-h shell-model calculations for inelastic pion scattering on ^{16}O /207/ are presented in Figure 41, showing the importance of T = 0 state excitations in the pion resonance region and below.

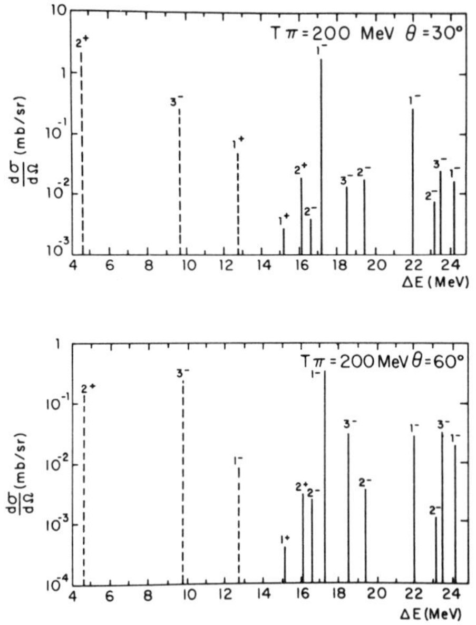

Fig. 42. Same as in Figure 41 for a C^{12} target /210/

Extensive shell-model calculations of pion scattering in standard DWIA approach have been carried out by LEE and TABAKIN /209/ in momentum space, and by HESS and EISENBERG /210/ with results of the latter shown in Figure 42. Due to the peaking at the nuclear surface of both the nuclear transition density and the pion wave function, it is found that pion-nucleus inelastic scattering near the resonance is mainly a surface reaction; the nuclear interior can only be explored by pions scattered at large angles, while at angles below the first diffraction minimum, the pion-nucleus inelastic scattering is insensitive to details of nuclear structure but just depends on the peak position of the transition density and its surface slope, see Figure 43. Unfortunately, at large angles the details of the reaction

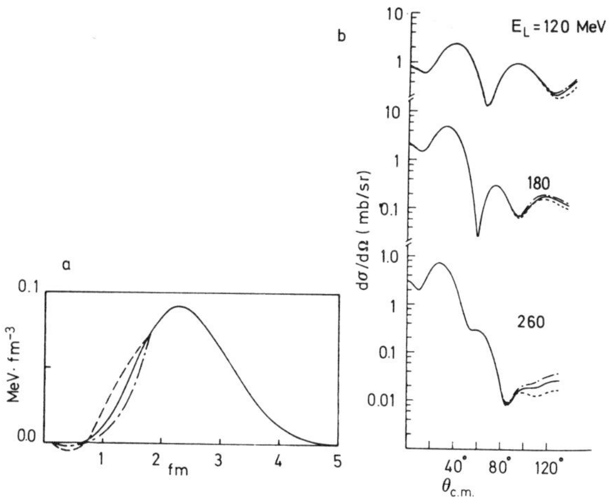

Fig. 43. a) The solid curve shows the 2^+ (4.43 MeV) transition charge density in ^{12}C. The dashed (dashed-dot) curve is the density generated by multiplying the collective 2^+ density for distances $r \lesssim a_0 = 1.8$ fm by $e^{-\beta(r-a_0)}$ ($e^{+\beta(r-a_0)}$), where $\beta = 0.36\ fm^{-1}$; b) differences generated in the corresponding inelastic (π, π') differential cross sections when calculated using the different definitions in a) /209/

mechanism become very important also, so that it is not completely clear how reliable the calculated DWIA result actually is. In addition, off-shell and relativistic effects might also be large /184/.

From a phenomenological point of view, momentum space codes for DWIA calculations seem to be suitable if one wants to utilize for the nuclear transition density the results of some direct parameterization of the corresponding electron scattering nuclear form factor.

Another advantage of the momentum space formulation is its flexibility in parameterizing the various off-shell dependencies. From a computational point of view, this may be important if one intends to specifically probe the off-shell sensitivity of π-nucleus scattering, as done, e.g., by LEE and TABAKIN /209/. Before, however, drawing any conclusions about physical results at this point, higher-order multiple scattering corrections should also be considered, and a better way should be found to treat genuine pion absorption by the nucleus.

Appendix A: Nonrelativistic Sum Rules in Nuclear Physics

Important sum rules for the analysis of electromagnetic transitions in nuclei have been derived by LEVINGER and BETHE /44/ and by GELL-MANN and TELEGDI /25/. These sum rules give the total strength of transitions of a particular multipole type from a given initial state to all other nuclear states combined. In the case where the initial state is the ground-state, such sum rules are directly related to the integrated cross sections for photon absorption, weighted by various powers of the photon energy. It was first realized by OPAT /211/ that, for example, in the case of electric dipole transitions in light nuclei it is possible to give sum rules for the strength of transitions from a given initial state to all states of a definite final spin and isospin.

As the electric dipole operator is a vector in configuration and in isospin space, three values of the final spin and three values of the final isospin are allowed in general, each with its own sum rule. This procedure might be employed to calculate isospin splitting, to which subject an extensive literature has been devoted (see also Sect. 1.4). A review of sum rules in connection with the role of isospin in electromagnetic transitions has been given in /212/, see also /50, 111, 112/. Here, we shall mainly discuss first-order energy-weighted sum rules which are dynamically of interest since they involve commutators of the nuclear Hamiltonian with the given excitation operator. The following results are useful in order to fill in the details in the developments of Sects. 1.3, 2.2 and 3.3, 4.4.

First-Order Energy-Weighted Sum Rule

Let us study the following sum rules:

$$\sum_f (E_f - E_0)\ |\langle f|\ \sum_{i=1}^{A} \tau_i^\alpha \exp(i\underline{q}\cdot\underline{r}_i)\ |0\rangle|^2 \qquad \text{(A,1a)}$$

$$\sum_f (E_f - E_0)\ |\langle f|\ \sum_{i=1}^{A} \tau_i^\alpha \exp(i\underline{q}\cdot\underline{r}_i)\ \underline{\sigma}_i\ |0\rangle|^2 . \qquad \text{(A,1b)}$$

We may use the following identity:

$$\frac{1}{2} \langle 0| \left[\mathcal{O}^+, \left[\mathcal{H}, \mathcal{O} \right] \right] |0\rangle = \sum_f (E_f - E_o) \, |\langle f| \, \mathcal{O} \, |0\rangle|^2 \tag{A,2a}$$

which holds if

$$|\langle f| \, \mathcal{O}^+ \, |0\rangle|^2 = |\langle f| \, \mathcal{O} \, |0\rangle|^2. \tag{A,2b}$$

In our case where

$$\mathcal{O} = \sum_i \tau_i^\alpha \, (\sigma_i) \, \exp(i\underline{q}\cdot\underline{r}_i) \, ,$$

plane waves do not introduce any additional complications since $|0\rangle$ and $|f\rangle$ are eigenstates of the parity operator, but we have to limit ourselves to $T_o = 0$ nuclei so that the condition (A,2b) is satisfied, since for $N > Z$ nuclei ($T_o \neq 0$), the operator τ^- only leads to $T_o + 1$ levels, while τ^+ excites $T_o - 1$, T_o and $T_o + 1$ levels. So, for $N > Z$ nuclei we cannot use the previous sum rule (A,2a), but we should use instead

$$\langle 0| \left[\mathcal{O}^+, \mathcal{H} \right] \mathcal{O} \, |0\rangle = \sum_f (E_f - E_o) \, \langle 0| \, \mathcal{O}^+ \, |f\rangle\langle f| \, \mathcal{O} \, |0\rangle$$

$$= \sum_f (E_f - E_o) \, |\langle 0| \, \mathcal{O} \, |f\rangle|^2. \tag{A,2c}$$

The left-hand side in general contains 3-body correlations while we shall see that (A,2a) contains only 2-body correlations. Here we study $T_o = 0$ nuclei only, and therefore may limit ourselves to $\tau^\alpha = \tau^3$.

Let us take a nuclear model Hamiltonian of the form

$$\mathcal{H} = \sum_{i=1}^{A} \frac{p_i^2}{2m} + \sum_{i<j} V(r_{ij}) + \xi \sum_i \underline{l}_i \cdot \underline{s}_i$$

$$+ \sum_{i<j} V^B(r_{ij}) \, P_{ij}^\sigma + \sum_{i<j} V^M(r_{ij}) \, P_{ij}^X - \sum_{i<j} V^H(r_{ij}) \, P_{ij}^\tau \tag{A,3}$$

$$+ \sum_{i<j} V_T(r_{ij}) \, S_{ij},$$

where ξ is the spin-orbit strength,

$$P_{ij}^\sigma = \frac{1 + \underline{\sigma}_i \cdot \underline{\sigma}_j}{2}$$

$$P_{ij}^\tau = \frac{1 + \underline{\tau}_i \cdot \underline{\tau}_j}{2}$$

P^X_{ij} = space exchange operator,

$$S_{ij} = \left\{ 3(\underline{\sigma}_i\cdot\underline{r}_{ij})(\underline{\sigma}_j\cdot\underline{r}_{ij})/r^2_{ij} - \underline{\sigma}_i\cdot\underline{\sigma}_j \right\}, \qquad \underline{r}_{ij} = \underline{r}_i - \underline{r}_j. \tag{A,4}$$

One then has

$$\left[\mathscr{H}, \textstyle\sum_i \tau^3_i \exp(i\underline{q}\cdot\underline{r}_i)\right] \quad = \text{kinetic energy contribution} +$$

$$+ \frac{\xi}{2}\sum_{i=1}^{A} (\underline{r}_i \times \underline{q}) \cdot \underline{\sigma}_i \exp(i\underline{q}\cdot\underline{r}_i)\,\tau^3_i + \tag{A,5a}$$

$$+ \sum_{i<j} \left[\exp(i\underline{q}\cdot\underline{r}_i) - \exp(i\underline{q}\cdot\underline{r}_j)\right] (\tau^3_j - \tau^3_i) \left[V^M(r_{ij})\,P^X_{ij} - V^H(r_{ij})\,P^\tau_{ij}\right],$$

$$\left[\mathscr{H}, \textstyle\sum_i \tau^3_i\underline{\sigma}_i \exp(i\underline{q}\cdot\underline{r}_i)\right] \quad = \text{kinetic energy contribution} +$$

$$+ i\xi \textstyle\sum_i \tau^3_i\underline{\sigma}_i\times\underline{l}_i \exp(i\underline{q}\cdot\underline{r}_i) + \frac{\xi}{2}\sum_i \underline{\sigma}_i\,(\underline{r}_i\times\underline{q}\cdot\underline{\sigma}_i) \exp(i\underline{q}\cdot\underline{r}_i)\,\tau^3_i$$

$$+ \sum_{i<j} \left[\exp(i\underline{q}\cdot\underline{r}_i)\,\tau^3_i - \exp(i\underline{q}\cdot\underline{r}_j)\,\tau^3_j\right] (\underline{\sigma}_j-\underline{\sigma}_i)\,V^B(r_{ij})\,P^\sigma_{ij} \tag{A,5b}$$

$$+ \sum_{i<j} \left[\exp(i\underline{q}\cdot\underline{r}_i) - \exp(i\underline{q}\cdot\underline{r}_j)\right] (\tau^3_j\,\underline{\sigma}_j - \tau^3_i\,\underline{\sigma}_i)\,P^X_{ij}\,V^M_{ij}$$

$$- \sum_{i<j} \left[\exp(i\underline{q}\cdot\underline{r}_i)\,\underline{\sigma}_i - \exp(i\underline{q}\cdot\underline{r}_j)\,\underline{\sigma}_j\right] (\tau^3_j - \tau^3_i)\,P^\tau_{ij}\,V^H_{ij}$$

$$- 4i \sum_{i<j} V_T(r_{ij})\,\tau^3_i \exp(i\underline{q}\cdot\underline{r}_i) \left\{3(\underline{\sigma}_j\cdot\underline{r}_{ij})(\underline{r}_{ij}\times\underline{\sigma}_i)/r^2_{ij} - \underline{\sigma}_j\times\underline{\sigma}_i\right\}.$$

Evaluating the double commutator, one obtains

$$\left[\textstyle\sum_i \tau^3_i\underline{\sigma}_i \exp(-i\underline{q}\cdot\underline{r}_i), \quad \left[\mathscr{H}, \sum_i \tau^3_i\underline{\sigma}_i \exp(i\underline{q}\cdot\underline{r}_i)\right]\right] =$$

$$= -2\,\xi \textstyle\sum_i \underline{l}_i\cdot\underline{\sigma}_i + \sum_{i<j} V_T(r_{ij}) \left\{16\,\underline{\sigma}_i\cdot\underline{\sigma}_j \left[1 - \tau^3_i\tau^3_j \exp i\underline{q}\cdot(\underline{r}_i-\underline{r}_j)\right]\right.$$

$$\left. - \frac{24}{r^2_{ij}} \left[2\,(\underline{\sigma}_j\cdot\underline{r}_{ij})(\underline{r}_{ij}\cdot\underline{\sigma}_i) + (\underline{\sigma}_j\times\underline{r}_{ij})\cdot(\underline{r}_{ij}\times\underline{\sigma}_i)\,\tau^3_i\tau^3_j \exp i\underline{q}(r_i-r_j)\right]\right\} \tag{A,6a}$$

$$+ \sum_{i<j} V^H(r_{ij}) \left|\exp(i\underline{q}\cdot\underline{r}_i)\,\underline{\sigma}_i - \exp(i\underline{q}\cdot\underline{r}_j)\,\underline{\sigma}_j\right|^2 (\tau^3_j-\tau^3_i)^2\,P^\tau_{ij}$$

$$- \sum_{i<j} V^B(r_{ij}) \left|\exp(i\underline{q}\cdot\underline{r}_i)\,\tau^3_i - \exp(i\underline{q}\cdot\underline{r}_j)\,\tau^3_j\right|^2 (\underline{\sigma}_j-\underline{\sigma}_i)^2\,P^\sigma_{ij}$$

$$- \sum_{i<j} V^M (r_{ij}) | \exp(i\underline{g}\cdot\underline{r}_i) - \exp(i\underline{g}\cdot\underline{r}_j) |^2 (\tau_j^3 \underline{\sigma}_j - \tau_i^3 \underline{\sigma}_i)^2 P_{ij}^X .$$

This result may be rewritten as

$$\begin{aligned} &- 4 \xi \sum_i \underline{l}_i \cdot \underline{s}_i - 8 \sum_{i<j} V_T (r_{ij}) S_{ij} (2 + \tau_j^3 \tau_i^3 \exp(i\underline{g}\cdot\underline{r}_{ij}) \\ &- 12 \sum_{i<j} V^B (r_{ij}) (1 - \frac{\underline{\sigma}_i \cdot \underline{\sigma}_j}{3}) (1 - \cos \underline{g}\cdot\underline{r}_{ij} \tau_i^3 \tau_j^3) P_{ij}^\sigma \\ &- 12 \sum_{i<j} V^M (r_{ij}) (1 - \cos \underline{g}\cdot\underline{r}_{ij}) (1 - \tau_i^3 \tau_j^3 \frac{\underline{\sigma}_i \cdot \underline{\sigma}_j}{3}) P_{ij}^X \\ &+ 12 \sum_{i<j} V^H (r_{ij}) (1 - \frac{\underline{\sigma}_i \cdot \underline{\sigma}_j}{3} \cos \underline{g}\cdot\underline{r}_{ij}) (1 - \tau_i^3 \tau_j^3) P_{ij}^\tau ; \end{aligned} \qquad (A,6b)$$

one finds similarly

$$\left[\sum_i \tau_i^3 \exp(-i\underline{g}\cdot\underline{r}_i), \left[\mathcal{H}, \sum_j \tau_j^3 \exp(i\underline{g}\cdot\underline{r}_j)\right]\right] = - \sum_{i<j} 4 (1 - \cos \underline{g}\cdot\underline{r}_{ij}) (1 - \tau_j^3 \tau_i^3) \times \left[P_{ij}^X V_{ij}^M - P_{ij}^\tau V_{ij}^H\right] . \qquad (A,7)$$

Expanding the foregoing results, it is possible to obtain various types of multipolar sum rules.

The procedure which we have outlined here may be formally generalized as done by NOBLE /213/ to yield a progenitor sum rule, i.e., a generating function which can be manipulated to yield a variety of sum rules. The important point to stress is that one must always start from commutation relations. Extensions to current algebra commutations relations will be studied in the Appendix B.

Appendix B: Exchange Effects and SU4 Invariance in Electromagnetic and Weak Transitions

1. Introduction

In the main text of this article, we have not given any explicit outline of how to deal with exchange effects as two-body corrections to the impulse approximation. We recall, however, that concerning the vector current, some two-body interaction effects have implicitly been included by the use of the continuity equation (6). In the following, we shall show that with the use of current algebra, one may go a certain way towards incorporating the action of exchange effects in the weak and electromagnetic transition matrix elements.

Many previous efforts have been made to eliminate the meson degrees of freedom from nuclear wave functions and electromagnetic and weak transition operators. Within the context of a meson theory the problem may be completely solved, and one may find equivalent transition operators whose matrix elements taken between eigenstates of the nucleonic Hamiltonian (involving only internucleon potentials) describe correctly the transition amplitudes /214/. Ambiguities may arise, however, because as noted by BELL /215/, correlation effects in the wave functions may by a unitary transformation be transformed into many-body terms in the operators and vice versa, so that effective interactions and correlation functions are not separately well-defined concepts.

We study many-body effects in electromagnetic and weak nuclear transitions in connection with current algebra, and assume SU4 invariance /33/ of the nuclear Hamiltonian. In essence, we try to find exact relations between matrix elements taken between physical nuclear states with mesonic effects included, of the physical electromagnetic and weak currents, by making use of their algebraic properties.

In Sect. 2, we shall briefly describe the commutators we use, which will be current-current and current-density commutators. In terms of these, we construct the SU4 algebra in the way suggested by RADICATI /35/ and illustrate its usefulness.

In Sect. 3, we study the application to muon capture, concerning ourselves mainly with the equality of the vector and axial matrix elements and the relation of the vector matrix elements, in the unretarded dipole approximation, to the photoabsorption cross section. This last point seems important, because it has been emphasized by GREEN /216/ that one point ignored by FOLDY and WALECKA /31/ in their treatment of muon capture on doubly closed shell nuclei is that of taking properly into account exchange effects, when they relate total muon capture rates to the photo-absorption cross section. While it is clear that exchange effects strongly affect the latter, it is not clear how they should appear in muon capture. Indeed in muon capture as far as the vector matrix elements are concerned, one deals with the density V_4 which is commonly accepted /27/ not to be as largely affected by mesonic effects as the current V_k that enters into the expression for the photo-absorption cross section. Here, instead of ($\underline{J}$, ρ) we use the notation J_μ with $\mu = 1...4$. This notation simplifies, e.g., the expression for the current algebra commutators. Furthermore, the coupling constants such as e and G_V, etc., are now taken out from the definition of the currents. The classical sum rule of (13), according to recent experimental data /118/, is indeed modified up to 100 %. In this connection,SIEGERT's theorem /27/ may be interpreted by stating that impulse approximation may be applied to V_4 but not to V_k /217/, so that

$$V_4(\underline{x},0) \cong \sum_i \frac{1+\tau_i^3}{2} \delta(\underline{x}-\underline{x}_i), \tag{B,1a}$$

$$V_k(\underline{x},0) \neq \frac{1}{2} \sum_i \left\{ \frac{1+\tau_i^3}{2} \frac{(\underline{p}_i)_k}{m}, \delta(\underline{x}-\underline{x}_i) \right\} . \tag{B,1b}$$

Furthermore, there is no direct connection between axial and vector exchange effects /218/, so that the equality between vector and axial matrix elements needs to be reinvestigated (indeed it appears that axial exchange effects are very small /218/).

In Sect. 4, we shall give some general arguments to show in which circumstances the SU4 predictions may be expected to be of sufficient accuracy, even if we know that this symmetry is broken.

2. SU4 and Current Algebra

The free quark model leads to the commutation relations at equal times /219/

$$\left[\int V_4^3(x)d\underline{x}, \int A_\mu^j(y)d\underline{y}\right]_{x_0=y_0} = - \varepsilon_{3jk} \int A_\mu^k(y)d\underline{y} \tag{B,2a}$$

$$\left[\int A_1^1(x)d\underline{x}, \int A_1^2(y)d\underline{y}\right]_{x_0=y_0} = \int V_4^3(x)d\underline{x} \tag{B,2b}$$

$$\left[\int A_k^\pm(x)d\underline{x}, \int V_4^3(y) \exp(i\underline{y}\cdot\underline{\nu})\, d\underline{y}\right]_{x_0=y_0} = \mp i \int A_k^\pm(y) \exp(i\underline{y}\cdot\underline{\nu})\, d\underline{y}.$$

Here, V and A denote the physical vector and axial vector hadronic currents, the upper index referring to isospin and the lower one to Lorentz space. The relations (B,2a,b) for 1, μ = 4, and (B,2c) for k = 4 can be proved without invoking the quark model /220/. In the expression

$$\left[\int A_k^{\pm}(x)d\underline{x},\ V_4^3(y)\right]_{x_0=y_0} = \mp\, i\, A_k^{\pm}(y) \qquad (B,3)$$

for k ≠ 4 we might expect to have some contribution from the so-called Schwinger terms, but it is generally thought that the once-integrated commutation relations are free from Schwinger terms /219, 220/.

WIGNER /33/ has proved that, if the nuclear Hamiltonian H_0 includes potentials of Wigner and Majorana type only, the operators

$$(T^o)^{\alpha} = \frac{1}{2}\sum_i \tau^{\alpha}(i) \qquad (B,4a)$$

$$(S^o)_{\lambda} = \frac{1}{2}\sum_i \sigma_{\lambda}(i) \qquad (B,4b)$$

$$(Y^o)_{\lambda}^{\alpha} = \frac{1}{2}\sum_i \sigma_{\lambda}(i)\, \tau^{\alpha}(i) \qquad (B,4c)$$

commute with H_0. If the nuclear forces are sufficiently short range and attractive in the relative S states, the ground-states of A = 4n nuclei, $|0\rangle$, are scalar super-multiplets such that

$$(T^o)^{\alpha}|0\rangle = (S^o)_{\lambda}|0\rangle = (Y^o)_{\lambda}^{\alpha}|0\rangle = 0. \qquad (B,5)$$

This idea can be extended to the case where the nucleus consists of nucleons and mesons or quarks; as far as isospin is concerned, this is essentially the philosophy underlying the conserved vector current (CVC) hypothesis /221/.

We now discuss the SU4 algebra generated by the operators containing the physical currents,

$$T^{\alpha} = \int V_4^{\alpha}(x)d\underline{x} \qquad (B,6a)$$

$$Y_k^{\alpha} = \int A_k^{\alpha}(x)d\underline{x} \qquad (B,6b)$$

$$S_k = \int A_k^{o}(x)d\underline{x}, \qquad (B,6c)$$

where k = 1,2,3 and $A_{\lambda}^{o}(x)$ is the isoscalar axial current density. We note that the partially conserved axial vector current (PCAC) hypthesis /222/ relates $d\int A_4^{\alpha}(x)d\underline{x}/dt$

to $\int \phi_\alpha(x)dx$, where $\phi^\alpha(x)$ is the pion field. Thus $\int A_4^\alpha(x)d\underline{x}$ is not a constant of motion, but nevertheless the current algebra of the free quark model guarantees that T^α, Y_k^α, S_k, /108/, satisfy the same commutation relations as $(T^o)^\alpha$, $(Y^o)_k^\alpha$, $(S^o)_k$. It should be noted that $A_\lambda^\alpha(x)$ includes what is usually called the induced pseudoscalar and the two-body Gamow-Teller exchange effects. The matrix elements of V_λ^α and A_λ^α are experimentally observable through the coupling of hadrons and leptons.

The reason why we are concerned with the algebra generated by T^α, Y_k^α, S_k stems from the fact that it is a direct extension of Wigner's supermultiplet theory /107/. In order to relate the generators defined by (B,6) to the quantities appearing in conventional nuclear physics, we must truncate the whole Hilbert space into a model subspace, in which there exist only nonrelativistic nucleons. In this Hilbert subspace, one has effective currents whose single body parts give

$$(T^1)^\alpha \cong (T^o)^\alpha \tag{B,7a}$$

$$(Y^1)_k^\alpha \cong (Y^o)_k^\alpha \tag{B,7b}$$

$$(S^1)_k \cong (S^o)_k . \tag{B,7c}$$

We have omitted the possible axial renormalization factor in (B,6) /107/. In Sect. 3 we shall discuss how this renormalization factor may be taken into account in (B,7b).

Let us now consider semileptonic processes in which the hadronic current J is coupled to the leptonic current. The hadronic structure enters in the expression (averaging over initial and summing over final spins) /223/

$$W_{\lambda\rho} = \overline{\sum} (2\pi)^4 \delta^4(p'-p-q) <p|\ J_\rho^+(0)\ |p'><p'|\ J_\lambda(0)\ |p>. \tag{B,8a}$$

This may be written as

$$\overline{\sum} \frac{2\pi}{V} \iint \exp(i\underline{q}\cdot(\underline{x}-\underline{x}') <p|\ J_\rho^+(x',0)\ |p'> \times \delta(E_{p'} - E_p - E_q) <p'|\ J_\lambda(\underline{x},0)\ |p>\ d\underline{x}\ d\underline{x}'. \tag{B,8b}$$

If we sum over all the final hadronic states, we obtain apart from constant factors

$$W_{\lambda\rho} \sim \iint \exp(i\underline{q}\cdot(\underline{x}-\underline{x}') <p|\ J_P^+(\underline{x}',0)\ \delta(H - E_p - E_q)\ J_\lambda(\underline{x},0)\ |p>\ d\underline{x}\ d\underline{x}'. \tag{B,8c}$$

If we particularize J to be V or A, we can define $W_{\lambda P}^{(V,V)}$ and $W_{\lambda P}^{(A,A)}$ in an obvious fashion.

It is then easy to verify that

$$W_{k\lambda}^{(A,A)} \sim \iint \exp(i\underline{q}\cdot(\underline{x}-\underline{x}'))\langle p|\ A_l^{+}(\underline{x}',0)\ \delta(H - E_p - E_q)\ A_k(\underline{x},0)\ |p\rangle \sim \delta_{kl}\ W_{44}^{(V,V)} \quad (B,8d)$$

(for k, l = 1,2,3) if we make use of the commutator Eq. (B,2c), when $|p\rangle$ is a scalar supermultiplet.

3. Muon Capture in Scalar Supermultiplets

If $|0\rangle$ is a scalar supermultiplet and the Hamiltonian H is SU4 conserving, then, using the commutator of (B,2c), we are led to

$$\begin{aligned}\textstyle\sum_f |\langle f|\int A_k^{\pm}(\underline{y},0)\ \exp(i\underline{y}\cdot\underline{\nu})\ d\underline{y}\ |0\rangle|^2 &= \\ = \textstyle\sum_f |\langle f|\int V_4^{3}(\underline{y},0)\ \exp(i\underline{y}\cdot\underline{\nu})\ d\underline{y}\ |0\rangle|^2 .\end{aligned} \quad (B,9)$$

From this equality, it follows that

$$M_V^2 = M_A^2, \quad (B,10)$$

where

$$M_V^2 = \sum_f \frac{\nu^2}{m_\mu^2} \int \frac{d\hat{\nu}}{4\pi}\ |\langle f|\int V_4^{\pm}(\underline{y},0)\ \exp(i\underline{y}\cdot\underline{\nu})\ d\underline{y}\ |0\rangle|^2 \quad (B,11a)$$

$$M_A^2 = \sum_f \frac{\nu^2}{m_\mu^2} \int \frac{d\hat{\nu}}{4\pi}\ |\langle f|\int A_k^{\pm}(\underline{y},0)\ \exp(i\underline{y}\cdot\underline{\nu})\ dy\ |0\rangle|^2 \quad (B,11b)$$

and $\nu = m_\mu - \varepsilon_\mu - (E_f - E_0)$, where ε_μ takes into account the binding energy of the muon, the difference in mass between neutron and proton, and the Coulomb displacement between analog states.

A proof of (B,10) is obtained by rewriting

$$\begin{aligned}M_V^2 = \textstyle\sum_f \int \frac{d\hat{\nu}}{4\pi}\ |(m_\mu-\varepsilon_\mu)\ \langle f|\ V_4^{\pm}(\underline{y},0)\ \exp(i\underline{y}\cdot\underline{\nu})\ d\underline{y}\ |0\rangle\ - \\ - \langle f|\left[H,\ \int V_4^{\pm}(\underline{y},0)\ \exp(i\underline{y}\cdot\underline{\nu})\ d\underline{y}\right]|0\rangle|^2\end{aligned} \quad (B,12)$$

(and similarly for M_A^2), and repeating the procedure which leads to (B,9).

The difference between the conventional theory for muon capture and the present theory lies in the fact that in the conventional theory, we have three matrix elements M_V^2, M_A^2, and M_P^2 (56), but in the present theory we have M_V^2 and M_A^2 only.

In the Wigner SU4 limit, the relation $M_V^2 = M_A^2 = M_P^2$ holds, and in the present case, we have $M_V^2 = M_A^2$.

Furthermore, in the SU4 limit of the usual theory the muon capture rate Λ is given by

$$\Lambda = \frac{m_\mu^2}{2\pi} |\phi|^2_{muon} \left\{G_V^2 + 3G_A^2 + G_P^2 - 2G_P G_A\right\} (M^o)_V^2 \qquad (B,13a)$$

[the matrix element M_V^2 of the conventional theory is now called $(M^o)_V^2$], where ϕ is the muon wave function, $G_V \cong 1.01G$, $G_A \cong -1.55G$, $G_P \sim 0.58G$ and $G(\cong 10^{-5} m^{-2})$ is the Fermi constant including the Cabibbo $\cos\theta_c$ factor /34/. In the present SU4 approximation, we have

$$\Lambda = \frac{m_\mu^2}{2\pi} |\phi|^2_{muon} G^2 4M_V^2 \qquad (B,13b)$$

in which M_V^2, defined as in (B,11a) depends on momentum transfer and contains, for example, magnetic contributions.

Equation (B,13b) appears rather unfamiliar, but it will now become clear how it can be related to (B,13a).

First of all, we have to obtain the renormalization constant relating $(Y^1)_k^\alpha$ to $(Y^o)_k^\alpha$. It is seen by taking the commutator of (B,2b) between nucleon states that the SU4 relation is broken even in the single nucleon system, which therefore does not belong to a pure representation of this group (this result was emphasized by RADICATI /35/). Therefore,

$$(Y^1)_k^\alpha = g_A (Y^o)_k^\alpha, \qquad (B,14)$$

where from beta decay experiments /34/, $g_A = 1.23$. Using this result, (B,13b) transforms into /224/

$$\Lambda = \frac{m_\mu^2}{2\pi} |\phi|^2_{muon} (1 + 3g_A^2) \cdot M_V^2. \qquad (B,15a)$$

Numerically, (B,13a) is equivalent to /224/

$$\Lambda = \frac{m_\mu^2}{2\pi} |\phi|^2_{muon} (1 + 3g_A^2) \cdot 1.08\, (M^o)_V^2. \qquad (B,15b)$$

Since in impulse approximation $M_V^2 = 1.2\ (M^o)_V^2$ /34/, the difference between (B,13a) and (B,15a) is not very large, even though these relations are derived under different hyptheses.

We now turn to the evaluation of M_V^2. Because of the continuity equation for the vector current, (6), we have

$$\int V_k(x)d\underline{x} = -\int x_k(\underline{\nabla}\cdot\underline{V}(x))d\underline{x} = \int x_k\, \dot{V}_4(x)d\underline{x} = i\left[H, \int x_k\, V_4(x)d\underline{x}\right]. \qquad (B,16)$$

Thus, in the unretarded dipole approximation M_V^2 may be related to $\int V_k(x)dx$, which in turn is related to the dipole photoabsorption cross section independently of the presence of exchange currents.

4. Discussion of SU4 Breaking Effects

Most of the previous arguments, which were based on exact SU4 symmetry, are valid also when instead of the commutation property

$$\left[H, Y_k^\alpha\right] = 0 \qquad (B,17)$$

we only require that

$$\left\{\left[H, Y_k^\alpha\right] - Y_k^\alpha\,\Delta\right\}|0> \cong 0, \qquad (B,18a)$$

where Δ is a constant defined by

$$\Delta \cong \frac{<0|(Y_k^\alpha)^+\left[H, Y_k^\alpha\right]|0>}{<0|(Y_k^\alpha)^+\ (Y_k^\alpha)|0>}. \qquad (B,18b)$$

Even if the commutator $[H, Y_k^\alpha]$ is not small, the SU4 relations are still valid to first order in the symmetry breaking if (B,18a) holds, with only minor modifications.

The history of the discovery of isobaric analogue states in the early 1960's tells us that /225/

$$\left[H, T^\alpha\right] \neq \text{small} \qquad (B,19a)$$

but

$$\left[H, T^\alpha\right] - T^\alpha\,\Delta = \text{small}, \qquad (B,19b)$$

Δ being a constant called the single-particle Coulomb displacement. If a similar relation is valid in our case, i.e., if (B,18) holds, we still obtain (B,9 and 10) without assuming any detailed knowledge of nuclear structure; for example, (B,10) was first found within the framework of the harmonic oscillator shell model /226/.

In order to prove the previous statements, let us study again the relation of (B,9), rewritten as

$$\begin{aligned}
&\textstyle\sum_f |\langle f| \int A_k^{\pm}(\underline{y},0)\, \exp(i\underline{y}\cdot\underline{\nu})\, d\underline{y}\, |0\rangle|^2 = \\
&\textstyle\sum_f |\langle f| \int V_4^{3}(\underline{y},0)\, \exp(i\underline{y}\cdot\underline{\nu})\, d\underline{y}\, |0\rangle|^2 \\
&+ \langle 0|\, Y_k^{\pm} \int A_k^{\mp}(\underline{y},0)\, \exp(i\underline{y}\cdot\underline{\nu}) \int V_4^{3}(\underline{y},0)\, \exp(i\underline{y}\cdot\underline{\nu})\, d\underline{y}\, |0\rangle \\
&- \langle 0| \int V_4^{3}(\underline{y},0)\, \exp(-i\underline{y}\cdot\underline{\nu})\, d\underline{y}\, Y_k^{\mp} \int V_4^{3}(\underline{y},0)\, \exp(i\underline{y}\cdot\underline{\nu})\, d\underline{y}\, Y_k^{\pm}\, |0\rangle \\
&+ \langle 0|\, Y_k^{\mp} \int V_4^{3}(\underline{y},0)\, \exp(-i\underline{y}\cdot\underline{\nu})\, d\underline{y} \int V_4^{3}(\underline{y},0)\, \exp(i\underline{y}\cdot\underline{\nu})\, d\underline{y}\, Y_k^{\pm}\, |0\rangle .
\end{aligned} \tag{B,20}$$

For the corrective terms, i.e., the last three terms of (B,20) which vanished in the previous case of (B,9), use has been made of the closure approximation. There are two kinds of such terms, namely i) terms without $A_k^{\mp}$, and ii) terms containing $A_k^{\mp}$. If first-order perturbation theory is valid, i.e.,

$$|0\rangle = |0\rangle_o + (1 - |0\rangle_o \langle 0|_o)\, \frac{1}{E_o - H_o}\, (1 - |0\rangle_o \langle 0|_o)\, H_1\, |0\rangle_o , \tag{B,21a}$$

where

$$Y_k^{\pm}\, |0\rangle_o = 0,\ \left[H_o, Y_k^{\pm}\right] = 0,\ \left[H_1\ Y_k^{\pm}\right] \neq 0 , \tag{B,21b}$$

then we can rewrite

$$\begin{aligned}
&\int V_4^{3}(\underline{y},0)\, \exp(i\underline{y}\cdot\underline{\nu})\, d\underline{y}\, Y_k^{\pm}\, |0\rangle \cong \int V_4^{3}(\underline{y},0)\, \exp(i\underline{y}\cdot\underline{\nu})\, d\underline{y}\ \cdot \\
&\cdot (1 - |0\rangle_o \langle 0|_o)\, \frac{1}{E_o - H_o}\,)\, (1 - |0\rangle_o \langle 0|_o) \left[Y_k^{\pm}, H_1\right] |0\rangle_o .
\end{aligned} \tag{B,22}$$

In a similar way, one may treat the quantity

$$\langle 0|\, Y_k^{\pm} \int A_k^{\mp}(\underline{y},0)\, \exp(-i\underline{y}\cdot\underline{\nu})\, d\underline{y} . \tag{B,23}$$

If now

$$\left[H_1, Y_k^{\pm}\right] = \Delta\, Y_k^{\pm} \tag{B,24}$$

which also leads to the assumption of (B,18a), then the terms i) and ii) clearly vanish.

Finally, let us consider the second term in (B,12). From the so-called AHRENS-FEENBERG approximation /227/

$$\left[H, \int V_4^{\pm}(\underline{y},0)\, \exp(i\underline{y}\cdot\underline{\nu})\, d\underline{y}\right] = \Delta E_F \int V_4^{\pm}(\underline{y},0)\, \exp(i\underline{y}\cdot\underline{\nu})\, d\underline{y}\,, \qquad (B,25a)$$

$$\left[H, \int A_k^{\pm}(\underline{y},0)\, \exp(i\underline{y}\cdot\underline{\nu})\, d\underline{y}\right] = \Delta E_{GT} \int A_k^{\pm}(\underline{y},0)\, \exp(i\underline{y}\cdot\underline{\nu})\, d\underline{y}. \qquad (B,25b)$$

We have if the dipole mode is assumed to be dominant

$$\left[H, \int V_4^{\pm}(\underline{y},0)(i\underline{y}\cdot\underline{\nu})\, d\underline{y}\right] \cong \Delta E_F \int V_4^{\pm}(\underline{y},0)(i\underline{y}\cdot\underline{\nu})\, d\underline{y}. \qquad (B,26)$$

Therefore we obtain

$$M_V^2/M_A^2 = (m_\mu - \varepsilon_\mu - \Delta E_F)^4/(m_\mu - \varepsilon_\mu - \Delta E_{GT})^4. \qquad (B,27)$$

(In this approximation the SU4 symmetry implies $\Delta E_F = \Delta E_{GT}$ whereas a breaking means $\Delta E_F \neq \Delta E_{GT}$).

The Ahrens-Feenberg approximation is justified when for an operator $\mathcal{O}$ the following relation holds:

$$\langle n|\left[H,\mathcal{O}\right]|0\rangle \cong \left\{\langle n|H|n\rangle - \langle 0|H|0\rangle\right\} \langle n|\,\mathcal{O}\,|0\rangle, \qquad (B,28)$$

where $|0\rangle$ and $|n\rangle$ are arbitrary states. Equation (B,28) holds if the nondiagonal matrix elements of H are negligible or, in other words, if nondiagonal matrix elements of the specific nuclear interaction are small between states with configuration differing in only a few orbitals. (Note, e.g., that a long-range exchange potential satisfies this condition /106/).

References

1 G.C. Baldwin, G.S. Klaiber: Phys. Rev. 71, 3 (1947); 73, 1156 (1948)
J.L. Lawson, M.L. Perlman: Phys. Rev. 74, 1190 (1948)
2 B.M. Spicer: Adv. Nucl. Phys. 2, 1 (1969)
3 E. Hayward: "Photonuclear Reactions", National Bureau of Standards, Washington, DC, NBS Monograph 118 (1970)
4 L.W. Fagg: Rev. Mod. Phys. 47, 683 (1975)
5 M.B. Lewis, F.E. Bertrand: Nucl. Phys. A196, 337 (1972)
6 G.R. Satchler: Comments Nucl. Part. Phys. 5, 145 (1972)
7 G.R. Satchler: Phys. Lett. 14C, 97 (1974)
8 F.E. Bertrand: Ann. Rev. Nucl. Sci. 26, 457 (1976)
9 G.A. Savitskij, A.A. Nemashkalo, V.M. Khyastunov: KFTI preprint, KFTI 76-16, Kharkov (1976)
10 Y. Torizuka: In Proc. Kawatabi Conf. on New Giant Resonances, Suppl. Res. Rept. Lab. Sci. Tohoku Univ. 8, 1 (1975)
11 R. Bergère, P. Carlos: "Les Resonances Géantes", Note CEA-N-1768, CEN Saclay (Jan. 1975)
12 R. Raphael, H. Überall, C. Werntz: Phys. Rev. 152, 899 (1966)
13 H. Überall: Electron Scattering from Complex Nuclei (Academic Press, New York, 1971)
14 Y. Torizuka, et al.: In Proc. Intern. Conf. on Nuclear Structure Using Electron Scattering and Photoreaction, Sendai, Japan 1972, ed. by K. Shoda, H. Ui (Tohoku Univ., Sendai 1972)
15 Ameenah R. Farhan, J. George, H. Überall: Nucl. Phys. A305, 189 (1978)
16 R. Silberberg: In Proc. 1976 DUMAND Summer Workshop, ed. by A. Roberts, Office of Publications, Fermi National Accelerator Lab., Batavia, IL, p. 55
17 A. Roberts, H. Blood, J. Learned, F. Reines: In Proc. Intern. Neutrino Conf. Aachen 1976, ed. by H. Faissner, H. Reithler, P. Zervas (F. Vieweg, Braunschweig 1977) p. 688
18 P. Alibran, et al.: Phys. Lett. 74B, 422 (1978)
19 H.V. Geramb, R. Sprickmann, G.L. Strobel: Nucl. Phys. A199, 545 (1973)
20 T. de Forest, J.D. Walecka: Adv. Phys. 15, 1 (1966)
21 H. Überall: In Springer Tracts in Modern Physics, Vol. 49 (Springer, Berlin, Heidelberg, New York 1969) p. 1
22 H. Morita: Beta Decay and Muon Capture (Benjamin, Reading, Mass. 1973)
23 H. Überall: In Springer Tracts in Modern Physics, Vol. 71 (Springer, Berlin, Heidelberg, New York 1974) p. 1
24 T.E.O. Ericson: Lecture notes, Herceg Novi Summer School (1969)
25 V.L. Telegdi, M. Gell-Mann: Phys. Rev. 91, 169 (1953)
26 E. Fermi: Ric. Sci. 7, 13 (1936)
27 A.J.F. Siegert: Phys. Rev. 52, 787 (1937)
28 W. Czyz, L. Lesniak, A. Malecki: Ann. Phys. (N.Y.) 42, 97 (1967)
29 A. Fujii, H. Primakoff: Nuovo Cimento 12, 327 (1959)
30 A.J. Armini, J.W. Sunier, R. Richardson: Phys. Rev. 165, 1194 (1968)
31 L. Foldy, J.D. Walecka: Nuovo Cimento 34, 1026 (1964)
32 J.D. Walecka: Preludes in Theoretical Physics in honour of V.F. Weisskopf, ed. by A. DeShalit, H. Feshbach, L. Van Hove (North-Holland, Amsterdam 1966)
33 E.P. Wigner: Phys. Rev. 51, 106 (1937)
34 F. Cannata, R. Graves, H. Überall: Riv. Nuovo Cim. 7, 133 (1977)

35 L.A. Radicati: Remarks on Wigner's Supermultiplet Theory on its 34th anniversary. Proc. 15th Solvay Meeting, Brussels (1970)
36 A.K. Petrauskas, V.V. Vanagas: Sov. J. Nucl. Phys. 8, 270 (1969)
37 N.C. Mukhopadhyay, F. Cannata: Phys. Lett. 51B, 225 (1974)
38 P.T. Nang: Nucl. Phys. A185, 413 (1972)
39 H. Überall: Acta Phys. Austriaca 30, 89 (1969)
40 H. Überall: Tech. Rpt. No. 70-035, Univ. of Maryland (1970)
41 H.W. Baer, K.M. Crowe, P. Truol: Adv. Nucl. Phys. 9, 177 (1977)
42 F. Cannata, H. Überall: Lett. Nuovo Cimento 10, 645 (1974)
43 W.R. Dodge, E. Hayward, R.G. Leich, B.H. Patrick, R. Starr: Proc. Giant Multipole Resonance Topical Conf., Oak Ridge, Tenn. (Oct. 1979)
44 J.S. Levinger, H.A. Bethe: Phys. Rev. 78, 115 (1950)
45 L.A. Radicati: Phys. Rev. 87, 521 (1952)
46 H. Überall: Suppl. Nuovo Cimento 4, 781 (1966)
47 M. Goldhaber, E. Teller: Phys. Rev. 74, 1046 (1948)
48 A. Faessler: Intern. Conf. Selected Topics in Nucl. Struct., Dubna, USSR, June 15-19 (1976)
49 T.W. Donnelly, J.D. Walecka: Ann. Rev. Nucl. Sci. 25, 329 (1975)
50 R. Leonardi, M. Rosa-Clot: Riv. Nuovo Cimento 1, 1 (1971)
51 H. Überall: Phys. Rev. 137, B 502 (1965)
52 W. Wild: Bayer. Akad. Wiss. Math. Naturwiss. Klasse 18, 371 (1956)
53 A.E. Glassgold, W. Heckrotte, K.M. Watson: Ann. Phys. (N.Y.) 6, 1 (1959)
54 S. Fujii: Progr. Theor. Phys. Suppl. 97 (1968)
55 H.M. Ferdinande, R.E. Van der Vyer: Nederl. Tijdschr. Natuurk. A44, 144 (1978)
56 H. Steinwedel, J.H. Jensen: Z. Naturforsch. A5, 413 (1950)
57 M. Danos: Nucl. Phys. 5, 23 (1958)
58 W.D. Myers, W. Swiatecki, T. Kodama, L.Y. El-Jaick, E.R. Hilf: Phys. Rev. C15, 2032 (1977)
59 R. Mohan, M. Danos, L.C. Biedenharn: Phys. Rev. C3, 1740 (1971)
60 M. Danos: Proc. Int. Conf. on Photonuclear Reactions and Applications, Lawrence Livermore Lab. (1973)
61 L.C. Biedenharn, M. Trivedi, M. Danos: Phys. Rev. C11, 1482 (1974)
62 A. Bohr, B.R. Mottelson: Nuclear Structure Vol. II (Benjamin, Reading, Mass. 1975)
63 J.W. Jury, B.L. Berman, D.D. Faul, P. Meyer, K.G. McNeill, J.G. Woodworth: Phys. Rev. C19, 1684 (1979)
R.G. Johnson, B.L. Berman, K.G. McNeill, J.C. Woodworth, J.W. Jury: Phys. Rev. C20, 27 (1979)
J.W. Jury, B.L. Berman, D.D. Faul, P. Meyer, J.G. Woodworth: Lawrence Livermore Lab. Preprint UCRL-83100, Aug. (1979)
64 D.J. Albert, A. Nagl, J. George, R.F. Wagner, H. Überall: Phys. Rev. C16, 503, 2452 (1977)
65 A. Yamaguchi, T. Terasawa, K. Nakahara, Y. Torizuka: Phys. Rev. C3, 1750 (1971)
66 H. Überall, B.A. Lamers, J.B. Langworthy, F.J. Kelly: Phys. Rev. C6, 1911 (1972)
67 T. deForest, J.D. Walecka, H. Van Praet, W.C. Barber: Phys. Lett. 16, 311 (1965)
68 W.C. Barber, F. Berthold, G. Fricke, F.E. Gudden: Phys. Rev. 120, 2081 (1960)
I.S. Gulkarov, N.G. Afanasyev, G.A. Savitsky, V.M. Khvastunov, N.G. Shevchenko: Phys. Lett. 27B, 417 (1968)
69 G.J. Van Praet: Nucl. Phys. 74, 219 (1965)
70 A. Hotta, K. Itoh, T. Saito: Phys. Rev. Lett. 33, 790 (1974)
71 F.H. Lewis Jr., J.D. Walecka: Phys. Rev. 133B, 849 (1964)
72 J.L. Friar: Phys. Rev. C1, 40 (1970)
73 F. Cannata, R. Leonardi: Nucl. Phys. A173, 665 (1971)
74 A. Migdal: J. Phys. USSR 8, 331 (1944)
75 M. Danos: Ann. Phys. (Leipzig) 10, 265 (1952)
76 T.W. Donnelly, G.E. Walker: Ann. Phys. (N.Y.) 60, 209 (1970)
77 R. Pitthan, Th. Walcher: Phys. Lett. 36B, 563 (1971); Z. Naturforsch. 27a, 1683 (1972)
78 S. Fukuda, Y. Torizuka: Phys. Rev. Lett. 29, 1109 (1972)
79 E. Hayward: Proc. Giant Multipole Resonance Topical Conf., Oak Ridge, Tenn. (Oct. 1979)

80 A. Bohr: In Nuclear Physics: An Int. Conf., ed. by R. Becker, C. Goodman, P. Stelson, A. Zucker (Academic Press, New York 1967)
B.R. Mottelson: In Proc. Int. Conf. Nucl. Structure, Kingston, 1960, ed. by D.A. Bromley, E.W. Vogt (Univ. of Toronto Press, Toronto/North Holland, Amsterdam 1960)
81 T. Suzuki: Nucl. Phys. A217, 182 (1973)
82 I. Hamamoto: Conf. on Nuclear Structure Studies Using Electron Scattering and Photoreaction, Sendai, 1972, Suppl. Res. Rep. Lab. Nucl. Sci., Tohoku Univ. 5, 208 (1972)
83 R. Pitthan: 11th Masurian Summer School in Nuclear Physics, Mikolajki, Masuria, Poland (Aug./Sept. 1978)
84 R. Bergère: In Photonuclear Reactions I, Lecture Notes in Physics, Vol. 61 (Springer, Berlin, Heidelberg, New York 1977) p. 1
85 Y. Torizuka: In "Colloque Franco-Japonais sur spectroscopie nucléaire et reaction nucléaire" Dogashima (1976)
86 Y. Torizuka: Proc. Meeting on Giant Resonances and Related Topics held at Institute for Nuclear Study, University of Tokyo, Tanashi-shi, Tokyo (1977)
87 Y. Torizuka: Proc. Sendai Conf. on Electro- and Photoexcitation, Sendai 1977, J. Phys. Soc. Jpn. 44, Suppl. 397-406 (1978)
88 J. Friederich: Nucl. Instrum. Methods 129, 505 (1075)
89 R.G. Satchler: Nucl. Phys. A195, 1 (1972)
90 K.W. McVoy: In Classical and Quantum Mechanical Aspects of Heavy Ion Collisions, Lecture Notes in Physics, Vol. 33 (Springer, Berlin, Heidelberg, New York 1975)
91 S.D. Drell, C.L. Schwartz: Phys. Rev. 112, 568 (1958)
92 K.W. McVoy, L. Van Hove: Phys. Rev. 125, 1034 (1962)
93 R.R. Betts, et al.: Phys. Rev. Lett. 39, 1183 (1977)
M. Buenerd, et al.: ibid 40, 1482 (1978)
P. Doll, et al.: ibid 42, 366 (1979)
94 H. Überall, A. Nagl, R.A. Lindgren, L.W. Fagg: Acta Phys. Austriaca 41, 341 (1975)
95 D. Kurath: Phys. Rev. 130, 1525 (1963)
96 J. Lichtenstadt, J. Heisenberg, C.N. Papanicolas, C.P. Sargent, A.N. Courtemanche, J.S. McCharthy: Phys. Rev. Lett. 40, 1127 (1978)
97 R.A. Lindgren, C.F. Williamson, S. Kowalski: Phys. Rev. Lett. 40, 504 (1978)
98 H. Zarek, B.O. Pich, T.E. Drake, D.J. Rowe, W. Bertozzi, C. Creswell, A. Hirsch, M.V. Hynes, S. Kowalski, B. Norum, F.N. Rad, C.P. Sargent, C.F. Williamson: Phys. Rev. Lett. 38, 750 (1977)
99 H. Überall: Notes of Lectures to the experimental group at Catholic University June (1977)
M. Traini: Phys. Rev. Lett. 41, 1535 (1978)
J.A. Montgomery, J.P. Ertel, H. Überall: To be published
100 A. Van der Woude: Proc. Giant Multipole resonance topical conference, Oak Ridge, Tenn.,October (1979)
101 E. Wolynec, W.R. Dodge, E. Hayward: Phys. Rev. Lett. 42, 27 (1979)
102 A. Johnston, T.E. Drake: J. Phys. A7, 898 (1974)
103 E.I. Kao, S. Fallieros: Phys. Rev. Lett. 25, 827 (1970)
T.J. Deal, S. Fallieros: Phys. Rev. C7, 1709 (1973)
T.J. Deal: Nucl. Phys. A217, 210 (1973)
104 Y. Torizuka, M. Oyamada, K. Nakahara, K. Sugiyama, Y. Kojima, T. Teresawa, K. Itoh, A. Yamaguchi, M. Kimura: Phys. Rev. Lett. 22, 544 (1969)
J.C. Bergstrom, W. Bertozzi, S. Kowalski, X.K. Maruyana, J.W. Lightbody Jr., S.P. Fivozinsky, S. Penner: Phys. Rev. Lett. 24, 152 (1970)
K. Itoh, M. Oyamada, Y. Torizuka: Phys. Rev. C6, 2181 (1970)
R.A. Eisenstein, D.W. Madsen, H. Theissen, L.S. Cardman, C.K. Bockelman: Phys. Rev. 188, 1815 (1969)
105 N.C. Mukhopadhyay: Phys. Lett. 30C, 1 (1977)
106 F. Cannata, R. Leonardi, M. Rosa-Clot: Phys. Lett. 32B, 6 (1970)
F. Cannata, J. Ros: Lett. Nuovo Cimento 8, 466 (1973)
107 F. Cannata, J. I. Fujita: Prog. Theor. Phys. 51, 811 (1974)
108 B. Lee: In High Energy Physics and Elementary Particles, IAEA Vienna (1965)
109 M. Fink, M. Gari, H. Hebach: Phys. Lett. 49B, 20 (1974)
110 A. Lodder, C.C. Jonker: Nucl. Phys. B2, 383 (1967)
111 S. Fallieros, B. Goulard: Nucl. Phys. 147A, 593 (1967)

112 E. Hayward, B.F. Gibson, J.S. O'Connell: Phys. Rev. C5, 846 (1972)
113 F. Cannata: Lett. Nuovo Cimento 13, 319 (1975)
114 H. Primakoff: Rev. Mod. Phys. 31, 802 (1959)
115 B. Goulard, H. Primakoff: Phys. Rev. C10, 2034 (1974)
116 R. Rosenfelder: Nucl. Phys. A290, 315 (1977)
117 R. Rosenfelder: Nucl. Phys. A298, 397 (1978)
118 J. Ahrens, et al.: Nucl. Phys. A251, 479 (1975)
119 M. Beiner, H. Flocard, N. Van Giai, P. Quentin: Nucl. Phys. A238, 29 (1975)
120 G.E. Walker: Phys. Rev. 174, 1290 (1968); Phys. Rev. C5, 1540 (1972)
R.J. McCarthy, G.E. Walker: Phys. Rev. C11, 383 (1975)
121 R.A. Eramzhyan, M. Gmitro, R.A. Sakaev, L.A. Tosjunjan: Nucl. Phys. A290, 294 (1977)
122 F. Tabakin: Ann. Phys. (N.Y.) 30, 51 (1964)
123 B. Pontecorvo: Proc. 1960 Annual Intern. Conf. on High Energy Physics, Rochester, N.Y., p. 617
Sov. Phys. Uspekhi 14, 235 (1971)
124 S. Weinberg: Phys. Rev. Lett. 19, 1264 (1967); 27, 1688 (1971); Phys. Rev. D5, 1412 (1972)
125 A. Salam, J.C. Ward: Phys. Lett. 13, 168 (1964)
126 W.B. Belyaev: Report N. 926, Dubna 1962 (unpublished)
127 C. Baltay: Comments Nucl. Part Phys. 8, 157 (1979)
128 F.J. Kelly, H. Überall: Phys. Rev. 158, 987 (1967)
129 T.W. Donnelly, J. Dubach, W.C. Haxton: Nucl. Phys. A251, 353 (1975)
130 R.W. King: Phys. Rev. 121, 1201 (1961)
131 S.S. Gershtein, N.V. Hieu, R.A. Eramzhyan: Sov. Phys. JETP 16, 1097 (1963)
132 N.A. Dadayan: Sov. J. Nucl. Phys. 24, 216 (1976)
133 S.M. Bilenkii, W.A. Dadayan: Sov. J. Nucl. Phys. 19, 459 (1974)
134 E.L. Fireman, H. Überall: Feasibility study for the Measurement of the Inelastic Neutrino Scattering Cross Section in ^{39}K, Research Proposal to Los Alamos Meson Physics Facility (Nov. 1975)
135 R. Davis: Private Communication
136 F. Cannata, R. Leonardi: Phys. Rev. C5, 1189 (1972)
137 S.L. Adler; Phys. Rev. 135B, 963 (1964)
138 H. Überall: Phys. Rev. 126, 876 (1962)
139 J.B. Langworthy, B.A. Lamers, H. Überall: Nucl. Phys. A280, 351 (1977)
140 M. Rosen, R. Raphael, H. Überall: Phys. Rev. 163, 927 (1967)
141 J.D. Murphy, H. Überall: Phys. Rev. C11, 829 (1975)
142 V. Devanathan, P.R. Subramanian, R.D. Graves, H. Überall: Phys. Lett. 57B, 241 (1975)
143 R.D. Graves, B.A. Lamers, A. Nagl, H. Überall, J.B. Langworthy, V. Devanathan, P.R. Subramanian: Canadian J. Phys. 58, 48 (1980)
144 A. Nagl, F. Cannata, H. Überall: Acta Phys. Austriaca 48, 267 (1978)
145 F.J. Kelly, H. Überall: Phys. Rev. C5, 1432 (1972)
146 J.S. O'Connell, T.W. Donnelly, J.D. Walecka: Phys. Rev. C6, 719 (1972)
147 T. deForest: Phys. Rev. 139, B1217 (1965)
148 V. Gillet, N. Vinh Mau: Nucl. Phys. 54, 321 (1964)
149 T.W. Donnelly, D. Hitlin, M. Schwartz, J.D. Walecka, S.J. Wiesner: Phys. Lett. 49B, 8 (1974)
150 V.N. Folomeshkin, S.S. Gershtein, M.Yu. Khlopov, M. Gmitro, R.A. Eramzhyan, L.A. Tosunjan: Inst. Nucl. Phys., Czech. Acad. of Sci Rpt. UJF-Th-58/75
151 G.V. Domogatsky, V.S. Imshennik, D.K. Nadyozhin: Neutrino '77 (Nauka, Moscow 1978)
152 N.M. Kroll, M.A. Ruderman: Phys. Rev. 93, 233 (1954)
153 A.I. Akhiezer, I.A. Akhiezer: Sov. J. Nucl. Phys. 8, 598 (1969)
154 F. Cannata, B.A. Lamers, C.W. Lucas, A. Nagl, H. Überall, C. Werntz, F.J. Kelly: Canadian J. Phys. 52, 1405 (1974)
155 T. Bressani: Riv. Nuovo Cimento 1, 268 (1971)
see also discussion remark, Proc. Int. Conf. High-Energy Phys. Nucl. Struct., Zurich 1977, ed. by M.P. Locher, p. 255
156 Y. Shamai, J. Alster, D. Ashery, S. Cochavi, M.A. Moinester, A.I. Yavin, E.D. Arthur, D.M. Drake: Phys. Rev. Lett. 36, 82 (1976)

157 M. Obu, T. Terasawa: Prog. Theor. Phys. 43, 1231 (1970)
158 P. Truöl: Meson-Nuclear Physics-1976 (Carnegie Mellon Conf.), AIP Conf. Proc. Series No. 23, p. 581
159 H.W. Baer: Proc. Int. Conf. High-Energy Phys. Nucl. Struct., Zurich 1977, ed. by M.P. Locher, p. 245
160 J.P. Perroud: Photopion Nuclear Physics, ed. by P. Stoler (Plenum Press, New York 1979) p. 69
161 H.R. Kissener, G.E. Dogotar, R.A. Eramzhyan, R.A. Sakaev: Nucl. Phys. A312, 394 (1978)
162 K. Ebert, J. Meyer-ter Vehn: Phys. Lett. 77B, 24 (1978)
163 J.C. Alder, et al: quoted in Ref. 159
164 J.A. Bistirlich, K.M. Crowe, A.S.L. Parsons, P. Skurek, P. Truöl: Phys. Rev. Lett. 25, 689 (1970)
165 J.D. Murphy, R. Raphael, H. Überall, R.F. Wagner, D.K. Anderson, C. Werntz: Phys. Rev. Lett. 19, 714 (1967)
166 F.J. Kelly, H. Überall: Nucl. Phys. A118, 302 (1968)
167 W.C. Lam, K. Gotow, B. MacDonald, W.P. Trower: Phys. Rev. C10, 72 (1974)
168 M. Kamimura, K. Ikeda, A. Arima: Nucl. Phys. A95, 129 (1967)
169 Y. Eisenberg, D. Kessler: Phys. Rev. 123, 1472 (1961)
170 D.K. Anderson, J.M. Eisenberg: Phys. Lett. 22, 164 (1966)
171 J.D. Vergados: Phys. Rev. C12, 1278 (1975)
172 E. Booth, H. Crannell, D. Sober: Private Communication
173 N. de Botton: Photopion Nucl. Phys., ed. by P. Stoler (Plenum Press, New York, 1979) p. 219
174 K. Shoda: Photopion Nucl. Phys., ed. by P. Stoler (Plenum Press, New York 1979) p. 175
175 E.C. Booth: Photopion Nucl. Phys., ed. by P. Stoler (Plenum Press, New York 1979) p. 129
176 W.M. MacDonald, E.T. Dressler, J.S. O'Connell: Phys. Rev. C19, 455 (1979)
177 F.J. Kelly, L.J. McDonald, H. Überall: Nucl. Phys. A139, 329 (1969)
178 F.A. Berends, A. Donnachie, D.L. Weaver: Nucl. Phys. B4, 1 (1967)
179 A. Nagl, H. Überall: Phys. Lett. 45B, 99 (1973)
180 A. Nagl, H. Überall: Phys. Lett. 63B, 291 (1976)
181 A. Nagl, H. Überall: Photopion Nucl. Phys., ed. by P. Stoler (Plenum Press, New York 1979) p. 155
182 F.L. Milder, E.C. Booth, B. Chasan, A.M. Bernstein, J. Comuzzi, G. Franklin, A. Nagl, H. Überall: Phys. Rev. C19, 1416 (1979)
183 M. Krell, T.E.O. Ericson: Nucl. Phys. B11,521 (1969)
184 J. Hüfner: Phys. Lett. 21C, 1 (1975)
185 M. Ericson, T.E.O. Ericson: Ann. Phys. (N.Y.) 36, 327 (1966)
186 G.E. Brown, B.K. Jennings, V.I. Rostokin: Phys. Lett. 50C, 227 (1979)
187 K. Shoda, H. Ohashi, K. Nakahara: Phys. Rev. Lett. 39, 1131 (1977)
188 J.B. Seaborn, V. Devanathan, H. Überall: Nucl. Phys. A219, 461 (1974)
189 J. Bernabeu, F. Cannata: Nucl. Phys. A215, 411 (1973)
190 R.A. Meyer, W.B. Walters, J.B. Hummel: Phys. Rev. 138, B1421 (1965)
191 B.N. Sung, K. Shoda, M. Yamazaki, K. Nakahara, H. Ohashi: Proc. Int. Conf. Nucl. Structure, Tokyo (1977)
192 K. Srinivasa Rao: Phys. Rev. C7, 1785 (1973)
193 J.A. Bistirlich, K.M. Crowe, A.S.L. Parsons, P. Skarek, P. Truöl: Phys. Rev. C5, 1867 (1972)
194 A. Nagl, F. Cannata, H. Überall: Phys. Rev. C12, 1586 (1975)
195 D.R. Bes, R.A. Broglia, B.S. Nilsson: Phys. Lett. 16C, 1 (1975)
196 W.G. Love, G.R. Satchler: Nucl. Phys. A101, 424 (1967)
197 P.J. Moffa, G.E. Walker: Nucl. Phys. A222, 140 (1974)
198 A. Picklesimer, G.E. Walker: Phys. Rev. C17, 237 (1978)
199 G.S. Adams, A.D. Bacher, G.T. Emery,W.P. Jones, R.T. Kouzes, D.W. Miller, A. Picklesimer, G.E. Walker: Phys. Rev. Lett. 38, 1387 (1977)
200 H.V. Geramb, K. Amos, R. Sprickmann, K.T. Knöpfle, M. Rogge, D. Ingham, C. Mayer-Böricke: Phys. Rev. C12, 1697 (1975)
201 G. Perrin, D. Lebrun, J. Chanvin, P. Martin, P. De Saintingnon, D. Eppel, H.V. Geramb, H.L. Yadav, V.A. Madsen: Phys. Lett. 68B, 55 (1977)
202 E.C. Halbert, J.B. McGrory, G.R. Satchler, J. Speth: Nucl. Phys. A245, 189 (1975)

203 A.M. Bernstein: Adv. Nucl. Phys. 3, 325 (1970)
A. Richter: Nuclear Physics with electromagnetic interactions, Lecture Notes in Physics, Vol. 108, ed. by H. Arenhövel, D. Drechsel, (Springer, Berlin, Heidelberg, New York 1979)
A.M. Bernstein, V.R. Brown, V.A. Madsen: Phys. Rev. Lett. 42, 425 (1979)
204 F.E. Bertrand, K. Van der Borg, A.G. Drentie, M.N. Harakek, J. Van der Plicht, A. Van der Woude: Phys. Rev. Lett. 40, 635 (1978)
205 D.H. Youngblood, C.M. Rosza, J.M. Moss, D.R. Brown, J.D. Bronson: Phys. Rev. Lett. 39, 1188 (1977)
206 F. Lenz: Proc. High Energy Physics and Nuclear Structure, Zürich 1977, ed. by M.P. Locher (Birkhäuser, Basel 1977) p. 175
207 M.K. Gupta, G.E. Walker: Nucl. Phys. A256, 444 (1976)
208 T. Bressani, F. Cannata: Lett. Nuovo Cimento 14, 523 (1975)
J. Arvieux, J.P. Albanese, M. Buenerd, D. Lebrun, E. Boschitz, C.H.Q. Ingram, J. Jansen: Phys. Rev. Lett. 42, 753 (1979)
209 T.S.H. Lee, F. Tabakin: Nucl. Phys. A226, 253 (1974)
210 A.T. Hess, J.M. Eisenberg: Nucl. Phys. A241, 493 (1975)
211 G.I. Opat: Nucl. Phys. 29, 486 (1962)
212 E.K. Warburton, J. Weneser: in Isospin in Nuclear Physics, ed. by D.H. Wilkinson (North-Holland, Amsterdam 1969)
213 J.V. Noble: Phys. Lett 40C, 243 (1978)
214 M. Chemtob: "Les Courants d'Interaction Nucléaires à Deux Corps", CEA Saclay Rpt. CEA-R-3960 and references quoted therein
215 J.S. Bell: In Proc. First Bergen International School of Physics, ed. by C. Fronsdal (Benjamin, New York 1962)
216 A.M. Green: In Proc. Spring School on Pion Interactions at Low and Medium Energies, CERN 71-14 (1971)
217 Y. Fujii, J. Fujita: Phys. Rev. 140, B239 (1965)
218 M. Chemtob, M. Rho: Nucl. Phys. A163, 1 (1971)
219 S.L. Adler, R.F. Dashen: Current Algebras (Benjamin, New York 1968)
220 J.J. Sakurai: Currents and Mesons (Univ. of Chicago Press, Chicago 1969)
221 R.P. Feynman, M. Gell-Mann: Phys. Rev. 109, 183 (1958)
222 M. Gell-Mann, M. Levy: Nuovo Cimento 16, 705 (1960)
223 S.D. Drell, J.D. Walecka: Ann. Phys. (N.Y.) 28, 18 (1964)
224 C.W. Kim, M. Ram: Phys. Rev. D1, 2651 (1970)
225 J. Fujita, K. Ikeda: Nucl. Phys. 67, 145 (1965)
226 J.R. Luyten, H.P.C. Rood, H.A. Tolhoek: Nucl. Phys. 41, 236 (1963)
227 T. Ahrens, E. Feenberg: Phys. Rev. 86, 64 (1952)

Subject Index

Nuclear Physics

1974. 116 figures. III, 245 pages
(Springer Tracts in Modern Physics, Volume 71)
ISBN 3-540-06641-1

Contents:
H. Überall: Study of Nuclear Structure by Muon Capture. – *P. Singer:* Emission of Particles Following Muon Capture in Intermediate and Heavy Nuclei. – *J. S. Levinger:* The Two- and Three-Body Problem.

S. Ferrara, R. Gatto, A. F. Grillo

Conformal Algebra in Space-Time and Operator Product Expansion

1973. II, 69 pages
(Springer Tracts in Modern Physics, Volume 67)
ISBN 3-540-06216-5

Contents:
Introduction to the Conformal Group in Space-Time. – Broken Conformal Symmetry. – Restrictions from Conformal Covariance on Equal-Time. Commutators. – Manifestly Conformal Covariant Structure of Space-Time. – Conformal Invariant Vacuum Expectation Values. – Operator Products and Conformal Invariance on the Light-Cone. – Consequences of Exact Conformal Symmetry on Operator Product Expansions. – Conclusions and Outlook.

Elementary Particle Physics

1976. 37 figures. VI, 145 pages
(Springer Tracts in Modern Physics, Volume 79)
ISBN 3-540-07778-2

Contents:
H. Rollnik, P. Stichel: Compton Scattering Compton Scattering in the Resonance Region. High-Energy Compton Scattering. – *E. Paul:* Status of Interference Experiments with Neutral Kaons: Interference Effects in a Beam of Coherent K_S^0 and K_L^0 and Possibilities of Measuring Them. K_S^0 Lifetime. (K_L^0-K_S^0) Mass Difference. Measurement of CP Violation in the Two-Pion Decay Modes. Search for CP Violation in Three-Pion Decay Modes. Test of the ΔS-ΔQ Rule in the Semileptonic Decay Modes. Analysis of the CP Violation Data Considering Unitarity. Possibilities of Explaining CP Violation.

Modern Three-Hadron Physics

Editor: A. W. Thomas

1977. 30 figures. XI, 250 pages
(Topics in Current Physics, Volume 2)
ISBN 3-540-07950-5

Contents:
I. R. Afnan, A. W. Thomas: Fundamentals of Three-Body Scattering Theory. – *L. R. Dodd:* Analytic Structure of On-Shell Three-Body Amplitudes. – *R. D. Amado:* Theory of Three-Body Final States. – *D. D. Brayshaw:* The Boundary Condition Method. – *R. Aaron:* A Relativistic Three-Body Theory. – *E. F. Redish:* Applications of Three-Body Methods to Many-Body Hadronic Systems.

Springer-Verlag Berlin Heidelberg New York